Geography, Health and Sustainability

With a global commitment to achieve gender equality by 2030, the SDGs present a historic opportunity to place gender as central to human progress across the globe. Gender equality, which requires the empowerment of all women and girls, is an explicit goal, in addition to being a fundamental prerequisite to and facilitator of most other SDGs. This edited collection provides a range of geographical and geospatial insights, from a variety of disciplinary and country-specific perspectives, to better understand gender and sustainable development. In addition to several African countries, Mexico, Japan, Canada, USA, and Cambodia are featured. A range of topical case studies examine women's domestic and care work, including water collection, breastfeeding, food purchasing, and caring for elderly family members. Access to healthcare services is examined in the case of breast screening and antenatal care. Women's engagement in the labor force is also addressed, with a specific look at the renewable energy sector; structural barriers to employment are discussed across a number of chapters, with clear strategies to break through these barriers. Finally, theoretical insights are proposed in better understanding and engaging in gendered inequalities in health.

Allison Williams is Professor in the School of Earth, Environment & Society at McMaster University. She is trained as a health geographer in quantitative, qualitative, and mixed-methods research. She leads a partnership grant to create carer-inclusive workplaces.

Isaac Luginaah is Professor in the Department of Geography and Environment at the University of Western Ontario and member of the College of The Royal Society of Canada. His research interests include environment and health, population health, and GIS applications in health.

Geographies of Health

Edited by Allison Williams, *Associate Professor, School of Geography and Earth Sciences, McMaster University, Canada,* and Susan Elliott, *Professor, Department of Geography and Environmental Management and School of Public Health and Health Systems, University of Waterloo, Canada*

There is growing interest in the geographies of health and a continued interest in what has more traditionally been labeled medical geography. The traditional focus of 'medical geography' on areas such as disease ecology, health service provision and disease mapping (all of which continue to reflect a mainly quantitative approach to inquiry) has evolved to a focus on a broader, theoretically informed epistemology of health geographies in an expanded international reach. As a result, we now find this subdiscipline characterized by a strongly theoretically-informed research agenda, embracing a range of methods (quantitative; qualitative and the integration of the two) of inquiry concerned with questions of: risk; representation and meaning; inequality and power; culture and difference, among others. Health mapping and modeling has simultaneously been strengthened by the technical advances made in multilevel modeling, advanced spatial analytic methods and GIS, while further engaging in questions related to health inequalities, population health and environmental degradation.

This series publishes superior quality research monographs and edited collections representing contemporary applications in the field; this encompasses original research as well as advances in methods, techniques and theories. The *Geographies of Health* series will capture the interest of a broad body of scholars, within the social sciences, the health sciences and beyond.

Public Health, Disease and Development in Africa
Edited by Ezekiel Kalipeni, Juliet Iwelunmor, Diana Grigsby-Toussaint and Imelda K. Moise

Blue Space, Health and Wellbeing
Hydrophilia Unbounded
Edited by Ronan Foley, Robin Kearns, Thomas Kistemann and Ben Wheeler

Geography, Health and Sustainability
Gender Matters Globally
Edited by Allison Williams and Isaac Luginaah

For a full list of titles in this series, please visit: www.routledge.com/Geographies-of-Health-Series/book-series/GHS

Geography, Health and Sustainability

Gender Matters Globally

Edited by Allison Williams and Isaac Luginaah

Routledge
Taylor & Francis Group
LONDON AND NEW YORK

First published 2022
by Routledge
2 Park Square, Milton Park, Abingdon, Oxon OX14 4RN

and by Routledge
605 Third Avenue, New York, NY 10158

Routledge is an imprint of the Taylor & Francis Group, an informa business

British Library Cataloguing-in-Publication Data
A catalogue record for this book is available from the British Library

Library of Congress Cataloging-in-Publication Data
A catalog record for this book has been requested

ISBN: 978-0-367-74390-1 (hbk)
ISBN: 978-0-367-74392-5 (pbk)
ISBN: 978-0-367-74391-8 (ebk)

DOI: 10.4324/9780367743918

Typeset in Times New Roman
by Apex CoVantage, LLC

Contents

Contributors

Meshack Achore is a Ph.D. student at the School of Kinesiology and Health Studies, Queen's University, Kingston, Canada. His doctoral research focuses on exploring how households in Sub-Sahara-Africa mobilize and use coping strategies to mitigate urban water insecurity, and the health impacts of these coping strategies.

Zahra Akbari is a human geographer, studying Ph.D. in Geography at Ferdowsi University of Mashhad. She received her master's degree from School of Geography and Earth Sciences at McMaster University. She has a bachelor's degree in Architecture. Zahra has mainly worked on immigrant carer-employees, home environment, Therapeutic landscape theory and Photovoice methodology.

Sophia Alhadeff holds a Bachelor of Geography with a concentration in Food, Agriculture and Society from Macalester College in St. Paul, Minnesota. Sophia's academic and professional interests include food justice, nutritional security, and embodiment geography.

Florence W. Anfaara, a Vanier Canada Scholarship winner, is pursuing a doctoral degree in the Department of Women's Studies and Feminist Research at the University of Western Ontario. Her research interests include the impact of infectious diseases on community health and well-being, transitional justice studies, women and gender in post-conflict societies.

George A. Atiim is a postdoctoral fellow with the United Nations University International Institute for Global Health (UNU-IIGH). His research broadly revolves around gender and health, the socio-cultural dimension of food allergies, food environments and the drivers of unhealthy diets, and participatory method in Sub-Saharan Africa.

Jemima Nomunume Baada is a Ph.D. candidate in the Department of Women's Studies and Feminist Research at Western University. Her research interests are in the areas of gender, migration, health, and climate change.

Bipasha Baruah holds the Canada Research Chair in Global Women's Issues at Western University. Her research aims to understand how to ensure that a global low-carbon economy will be more socially just than its fossil-fuel-based predecessor.

Elijah Bisung, Ph.D., is Assistant Professor in the School of Kinesiology and Health Studies, Queen's University, Kingston ON, Canada. He is a health geographer whose primary area of research focuses on social and environmental production of health and well-being.

Dr. Sheila A. Boamah is Assistant Professor at the School of Nursing at McMaster University. Her research interests include health system transformation and innovation, and quality and outcomes of underserved populations. Her research primarily investigates how clinical microsystems, technology, and organizational processes affect outcomes of care and quality of life.

Dr. Godfred O. Boateng is Assistant Professor of Global Health at The University of Texas at Arlington. Dr. Boateng is an expert in the design and application of culturally relevant scalable methodologies in understanding the multidimensional factors and processes that shape health and health equity across spatial scales and how these factors can be addressed in a sustained manner.

Rhonda A. Boateng holds an M.Sc. in Global Health and is pursuing a Ph.D. in Health Policy, Management and Evaluation. Data Science, Mental Health, medical education, and non-communicable diseases are among her research interests. She is a project coordinator within Education Services at the Centre for Addiction and Mental Health (Toronto).

Cristina D'Alessandro, Ph.D., is Associate Professor at Sciences-Po Paris Executive Education (France) and a member of the Centre of Governance, University of Ottawa (Canada). Economic and political geographer with 20+ years of teaching and research experience in prestigious institutions, she serves as an international consultant with various organizations and institutions.

Claudia Tello de la Torre is a Catedra-CONACYT researcher associated with CentroGeo. She holds a doctorate in economics from the University of Barcelona and master's degree in government and Public Affairs from the Latin American Faculty of Social Sciences (FLACSO-Mexico). Her research includes urban, regional, and metropolitan economic, spatial analysis, and applied econometrics.

Florence Dery is a Ph.D. student at the School of Kinesiology and Health Studies, Queen's University, Kingston, Canada. Her research broadly focuses on the water-health nexus, gender empowerment, and global health. She works on topics at the intersection of environment, development, and human well-being.

Margarita Parás Fernández is a Titular Technologist Research Professor "C" in the area of Geopolitics and Territory associated with CentroGeo.

Rabia Ferroukhi is the director of the International Renewable Energy Agency's (IRENA) Knowledge, Policy and Finance Centre. Dr. Ferroukhi brought to this position over 25 years of experience in the fields of energy, development, and environment.

Reiko Gotoh is Professor at Ibaraki University in Mito, Japan. She has engaged in interdisciplinary research projects on informatics, labor market policies, and gender equality. Her research primarily concerns policy evaluation and policy management, with particularly focusing on quantitative analysis of local government policies.

Paulina Grobet Vallarta is currently the coordinator of the Global Centre of Excellence on Gender Statistics (CEGS). She holds a master's degree in Demography by El Colegio de México, and a Sociology bachelor at the UAM. She is a Ph.D. candidate in Sociology, with a specialization in Population, from the University of Texas at Austin.

Yoshitaka Hojyo joined the City Government of Mito, Japan, in 1997. He has assumed since 2017 a function of Manager for Information Policy Section, Public Office of the Mayer, which oversees various information systems of the city and conducts public surveys including a Population and Housing Census.

Miya Ishitsuka is the manager of the Gender Equality Section, Department of Civic Cooperation, the City Government of Mito, Japan, since 2016. In this role, she is responsible for formulating and facilitating the implementation of official plans to promote gender equality and active participation of women in the city.

GurvirKalsi is a master's student at the School of Rehabilitation Science at Queen's University. He is broadly interested in the health and well-being of marginalized populations.

Joseph Kangmennaang, Ph.D., is Assistant Professor in School of Kinesiology and Health Studies, Queen's University, Kingston, ON. His research interests include global change, health, and well-being. He examines the relationships between inequality, health and well-being in the context of environmental, economic, and epidemiological changes in low- to middle-income countries.

Erica S. Lawson is Associate Professor in the Department of Women's Studies and Feminist Research at The University of Western Ontario. Her research focuses on the politics and practices of Black women's maternal activism for social and political change.

Celia García-Baños López is an associate programme officer in IRENA's Knowledge, Policy and Finance Centre in Abu Dhabi. She works on a diverse range of topics, including policy and socio-economic benefits of renewable energy.

William G. Moseley is DeWitt Wallace Professor of Geography, and director of the Food, Agriculture & Society Program, at Macalester College in Saint Paul, MN, USA. His research interests include tropical agriculture, food security, and development policy.

Keiko Osaki-Tomita is the President of Tokiwa University and Tokiwa Junior College, located in Mito, Japan. Prior to the current position, she had 30 years' professional career with the United Nations, responsible for research on population, international migration, gender and social integration, as well as the development of statistical systems.

Ola Osman is a Gates-Cambridge Scholar who is currently pursuing a doctoral degree at the University of Cambridge's Center for Multi-Disciplinary Gender Studies. Her work primarily focuses on mapping the continuities between racial slavery, its attendant gendered logics, and the Liberian civil war.

Yaa Adobea Owusu is Associate Professor, Head of Social Division, Institute of Statistical, Social, and Economic Research (ISSER), College of Humanities, University of Ghana, Legon. Her specializations include social-behavioral health, monitoring and evaluation, health services research, women's reproductive health, social and behavioral theory, and health communication/health education materials development.

Andrea Rishworth is a postdoctoral fellow in the Department of Geography, Geomatics and Environment, University of Toronto Mississauga. She is trained as a human geographer. Her research interests include aging, gender and care, global health disparities, and integrated knowledge translation.

Mengieng Ung has a background in environmental science and health geography with research interests in global health, health literacy and policy, climate change, sustainability, and GIS. Her work addresses communicable diseases (HIV/AIDS, TB), sexual and reproductive health, non-communicable diseases, dementia and mental well-being of the elderly, and healthy city projects.

Allison Williams is a Professor in the School of Earth, Environment & Society at McMaster University. She is trained as a health geographer in quantitative, qualitative, and mixed-methods research. She engages in social justice research to inform policy and program change. She is leading a partnership grant to create carer-inclusive workplaces.

1 Gender matters globally

Geography, health, and sustainability

Isaac Luginaah, Allison Williams, and Andrea Rishworth

Introduction

The intention of this edited collection is to bring attention to the centrality of ***geography*** and ***gender*** in reaching the 2030 Sustainability Development Goals (SDGs). The SDGs are made up of 17 goals, 169 targets, and 232 indicators, with 54 indicators being gender-specific (UN Women, 2018). With a global commitment to achieve gender equality by 2030, the SDGs present a historic opportunity to place gender as central to human progress across the globe. *Gender Equality*, which requires the empowerment of all women and girls, is an explicit goal, in addition to being a fundamental prerequisite to and facilitator of sustainable development. Achieving gender equality will translate into a range of sustainable development outcomes, including the building of peaceful, just and inclusive societies, through to ending hunger and poverty, achieving education for all, promoting prosperity, and protection of the planet and its natural resources (UN Women, 2018). This edited collection provides a range of geographical and geospatial insights, from a variety of disciplinary and country-specific perspectives, to better explain gender and sustainable development.

Why gender matters globally

Gender refers to the socially constructed characteristics and roles of women and men, in all their diversity, while sex refers to purely biological differences. The inherent biological differences between sexes form the basis of social norms that define what can be viewed as appropriate behavior for women and men while determining the differential social, economic, and political power between the sexes (Ridgeway, 2011). For example, in Sierra Leonne, social practices of female genital mutilation (FMG)[1] among the Bondo provide an acceptable way for females to gain access to womanhood and be rewarded with celebrations, gifts, and public recognition (Ameyaw et al., 2020). Although FMG undermines human rights legislation and raises long-term health concerns for women, this customary practice provides a means for females to gain critical social, economic, and political opportunities requisite for a flourishing life. Invariably, gender has remained a fundamental marker of social and economic stratification linked to exclusion,

DOI: 10.4324/9780367743918-1

especially in many developing countries. Existing evidence from sub-Saharan Africa indicates that, on average, resources are unevenly distributed within households with boys benefiting more than girls from investments in childcare, healthcare, and private education (UN, 2015). This means that women in the region are not only more likely to be undernourished and reside in poorer households compared to men (Brown et al., 2017), but also do three times the amount of unpaid care and domestic work than their male counterparts (UN Women, 2019).

Consequently, the behavior, attitudes, values, and beliefs that a particular sociocultural group considers appropriate for males and females connote their conception of gender (Butler, 2011). According to Butler (2011), gender roles are fluid and can shift over space and time. For instance, although many Latin American and Caribbean countries have made important progress in increasing the use of modern contraceptives, important spatial inequalities remain between countries in the region, with the lowest rates of contraception in Haiti and Bolivia, and the highest rates in Columbia, Costa Rica, and Paraguay. These disparities vary further within countries by wealth, education, and geography (de Leon et al., 2019). Spatiotemporal variations further structure the access and opportunities for males and females. For example, since the 1995 Beijing Platform for Action, more than two-thirds of all countries globally have reached gender parity[2] in enrollment in primary education and closing the gender gap in enrollment, but in countries that have not reached parity, particularly in Africa, the Middle East, and South Asia, girls are more likely to be disadvantaged than boys (UNESCO, 2019).

The underlying complexities of contextual and compositional factors that influence social relationships tend to shape our understanding of what it means to be male and female both individually and collectively (Risman et al., 2012). There have been persistent reports of the systematic gender differences in health and material well-being depending on one's social class, and the degree of inequality varies across countries and over time (Meinzen-Dick et al., 2014). The pervasiveness of global gender inequalities is happening in a context where males on average are better positioned in social, economic, and political hierarchies (Meinzen-Dick et al., 2014; Ridgeway, 2011; Carr, 2005). Moving away from the rather narrow understanding of gender equality as concerning only the differences between women and men, gender must be understood within the complexities of specific local contexts (Liverman, 2018a). By capturing the different experiences of men and women, gender can be understood as dynamic and layered with a range of multiple, intersecting social determinants that impact health. In Bangladesh, for example, access to potable water is as much a function of aquifer infrastructure as it is a function of gender, class, and position within family and social systems. While women in Bangladesh experience higher rates of water insecurity than men, inequalities in access and use of water resources exist and vary within the female population. Within hierarchical family structures, women typically command less access to cash, food, decision-making powers, education, and other livelihood resources compared to their household counterparts. Yet, despite inter- and intra-household gender inequality, when examining variations between women, those from households of higher socio-economic status are able to yield greater access to water resources, underscoring how intersecting positionalities are critical in the ways that gender relations and rights unfold in society, especially with respect to water (Sultana, 2011).

Globally, gender equity discussions are grounded in human rights principles and centered on the concept of universality (WHO, 2017). In this regard, attending to gender inequalities and their spatial and place specific dimensions requires concrete actions to eliminate inequalities in policies and practices. The systematic disadvantage and discrimination women face can be framed in terms of capabilities (as defined later), access to resources, and security dimensions (Meinzen-Dick et al., 2014; Agarwal, 2018). Capabilities are basic human abilities as measured by education, health, and nutrition (Sen, 1990; Anand et al., 2009). These are fundamental to individual well-being and are the means through which individuals can function and access other forms of well-being. Capabilities facilitate what an individual can do, are able to do, and dictate the protection of central freedoms that make for a good life (Sen, 1982; Nussbaum, 2008). For example, violence against women in places such as Libya and Tunisia are considered a capability of deprivation from bodily integrity whereby women are unable to move freely from place to place, and bodily boundaries are deemed unsovereign. That is, women do not have the capability to secure against assault, including sexual assault, child sexual abuse, and domestic violence or choice in opportunities for sexual satisfaction and for choice in matters of reproduction (Pyles, 2008). This is crucial, as bodily integrity is an important freedom in its own right as well as a means to further freedoms and economic opportunities. According to the capabilities approach, the government, via its social policies, is ultimately responsible for delivering the foundations of these capabilities (Nussbaum, 2008).

In relation to gender, there is differential access to both economic and natural resources by men and women. Access to resources is frequently influenced by equality in the opportunity to use or apply basic capabilities through access to economic assets (such as land and property) and resources (such as education, income, and employment). Without access to resources, women are unable to use their capabilities to ensure their well-being and the well-being of their families, communities, and societies (see Agarwal, 2018). In keeping with the population health promotion model, where sex and gender are recognized as social determinants of health (Phillips, 2005), feminist theories recognize gender as a prime component in all social interactions, and a determinant in accessing resources (Rocheleau, 1995). Underlying cultural norms around gender determine not only who is able to participate in the waged economy but also what types of labor are considered to be of economic value. For instance, the relations of power between men and women, especially in a patriarchal society, may create asymmetry in inherent capabilities, which in turn creates unequal access to and utilization of resources outside of the household (Sen, 1990). This is true in the patriarchal context of Tanzania, where high rates of gender inequality shape the realities and opportunities available to men and women. In Tanzania for instance, while social norms that deem men the household heads and main family breadwinners, and women are responsible for family matters and the dependence on men continue to persist, the impacts of even though structural adjustment reforms aimed to provide opportunities for women to seek paid employment and assume new roles as the household head. Yet, as women begin to take on greater financial responsibilities, men can develop hostility toward women's economic activity, and in so doing, reshape ideals of masculinity that revolve around abandonment of women, extra-marital relationships, excessive drinking, aggression, and violence toward women (Vyas & Jansen, 2018). An important point from this example is that changing social, economic, and gendered realities of men and women must be theorized as products of broader structural systems – political (e.g., lack of gender responsive policy-making), economic (e.g., uneven access to education, economic resources, employment opportunities), and legal (e.g., inadequate provision of legal and social services) – that reinforce the often disadvantaged status of women at both the individual and community scales (Montesanti, 2015).

Access and control of resources are negotiated between individuals from relative "bargaining positions," which are informed by a range of characteristics, such as the individual's perceived contribution to household livelihoods or the individual's social and economic position independent of the other household members (Sen, 1990). For example, in Ghana's Volta Region, the health seeking behavior for children is heavily tied to intra-household bargaining over access to and control over resources within the household (Tolhurst et al., 2008). While women are largely responsible for identifying illnesses and deciding which treatment to give children within the home, they are expected to involve their husbands in decisions over taking a child for care. Yet, since women have lower access to productive resources than their male counterparts, retain less control over jointly earned income, and are expected to carry out more non-income-earning activities within the household, the decision to seek care is heavily circumscribed by the competing bargaining power of women and men within the household.

Concerning women, such negotiation can only take place when they feel empowered in their roles (Boateng et al., 2014). With regard to the access and utilization of healthcare services, while there are obvious biological differences between men and women that shape the need for health care, more importantly the gendered social relations hold huge implications for health outcomes (Doyal, 2000; Moss, 2002). Existing work has extensively demonstrated that women have greater need for health services, as their relative material deprivation translates into poorer access and utilization when compared to their male counterparts (Dixon et al., 2011, 2014). Thus, the social relations of gender have material implications for the impact on health and, presumably, the demand for health insurance. Security within capabilities framing refers to a reduced vulnerability to gender-based violence and conflict, which can lead to physical and psychological harm and constrict individuals, households, and communities from fulfilling their potential (Nussbaum, 2008; Sen, 1990). The capabilities perspective recognizes the importance of understanding the experiences of individuals, and thus, provides a useful way for developing policies that can facilitate their overall well-being (Nussbaum, 2008). The World Economic Forum (2020) published the Global Gender Gap Index (GGGI), which examines gender inequality across the typical four key areas: health, education, economy, and politics. The 2020 report, like previous reports (Kim & Seo, 2017; Choe et al., 2017; Sharma et al., 2021), shows that although many countries have made progress since 2015 in reducing the gender gap, in some countries, the gender gap has widened, especially in terms of economic participation and governance opportunities. Persistent global gender inequalities means that we need to reinforce our understanding of gender as a social and relational process, rather than simply emphasizing the difference between women and men. An expanded understanding of gender, as shaped by economic, political, and cultural relationships, provides a novel starting point for progress on sustainable development that has yet to be fully and comprehensively realized (Meinzen-Dick et al., 2014). In response to persistent gender inequalities, governments have pledged to take further concrete actions to achieve gender equality, including strengthening the implementation of laws, policies, strategies, and institutional mechanisms, as well as work toward transforming discriminatory gender norms and practices. For instance, in 2014, the Government of Tajikistan began a process of advocacy and dialogue to identify the needs of women entrepreneurs and discuss initiatives to support them. In so doing, a *Taskforce to Support Women Entrepreneurship* was created to bring together representatives from the private sector, government, and the donor community in order to serve as a forum to create policies and reforms to support women's entrepreneurship (Oxfam, 2017).

In Indonesia, the government enacted *Law No. 23 Regarding Elimination of Violence in the Household* in order to provide domestic workers with access to legal protection, assistance, and redress violence committed against them in the course of their work, and in so doing, classify violence as an unacceptable human rights violation (Presiden Republik Indonesia, 2004). While these examples are important first steps to achieve gender equality, critical gaps remain to gender equalities *global* realization.

Gender and Sustainable Development Goals

The Millennium Declaration was signed by 189 countries in September 2000, when the Millennium Development Goals (MDGs) (UNDP, 2003) were adopted. The countries that signed this declaration committed themselves to achieving a concrete set of eight measurable goals by 2015. In the signatory countries, the MDGs aimed to address extreme poverty and its many dimensions, including income poverty, hunger, disease, lack of adequate shelter and exclusion, while promoting gender equality, education, and environmental sustainability (UNDP, 2003). The MDGs also included basic human rights, by emphasizing the rights of each person on the planet to health, education, shelter, and security. On the completion of 15 years of MDGs, while some significant gains were made regarding several MDG targets worldwide, important geographical variation in progress across regions and countries was also noted in the global report (UNDP, 2015). The general consensus was that millions of people are being left behind (especially from among the poorest and disadvantaged sections of society) because of their gender, sex, age, disability, ethnicity, or geographic and other social locations. These findings further corroborated the concerns raised by women's groups regarding the persistence of gender gaps that require more systemic attention. The failure by many countries to achieve the MDGs paved the way for UN Sustainable Development Goals (SDGs) in September 2015 (UN Women, 2018).

The SDGs were therefore "a universal call to action to end poverty, protect the planet, and ensure that all people enjoy peace and prosperity" with a motto, "Leave No One Behind" (UN Women, 2018). The SDGs combine economic, social, and environmental dimensions, with a particular emphasis on social inclusion. The 17 Goals and 160 Targets are to be achieved by 2030 (see Figure 1.1). The 17 goals are interconnected, which means that gains in any one area would catalyze achievements in others, with the potential to create greater synergies and impact. The SDGs underline the areas where states have failed to meet their gender-equality obligations. In particular, SDGs' Goal 5, "Achieve Gender Equality and Empower All Women and Girls," (UN Women, 2018) is central to the achievement of the other 16 goals. According to Agarwal (2018), the SDGs highlight women's human rights and gender equality, being universally endorsed by governments, donors, women's movements, civil society, and other international and national stakeholders. However, persistently high gender gaps remain in many countries around the world, whether they are developed or developing (Kim, 2017; Agarwal, 2018).

Figure 1.1 Social Development for Sustainable Development

Source: United Nations. Department of Economic and Social Affairs (Accessed 5 February 2021) © UNITED NATIONS

Copyright: www.un.org/development/desa/dspd/2030agenda-sdgs.html

In order to deal with the underlying causes of deep-rooted and persistent gender inequality, we must understand and implement culturally nuanced and contextualized approaches to achieving the SDGs (Song & Kim, 2013). Gender inequality is simply not acceptable in the world where half of the world's citizens are not provided with the same rights as the other half (Azcona & Bhatt, 2020; Kim, 2017). According to Esquivel and Sweetman (2016), the notion of "Leave No-one Behind" highlights the fact that the issues facing women in poverty do not arise from gender inequality only; rather, they are at the intersection of different dimensions of inequality, including race and class. Since the inception of the SDGs agenda, these differing norms and inequalities persist. Overall, most national systems still typically favor males, giving them more access than females to the capabilities, resources, and opportunities that are important for economic empowerment, political power, and social well-being (Esquivel & Sweetman, 2016; Meinzen-Dick et al., 2014; Ridgeway, 2011).

Goal 5 is based on a strong gender analysis, which understands gender inequality to possess a range of interconnected economic, political, and social characteristics (see Ritchie et al., 2018). In particular, SDG Goal 5 encompasses a multi-dimensional approach to gender equality with a wide range of targets that include (see Figure 1.2): end discrimination against women and girls (5.1); end all violence against and exploitation of women and girls (5.2); eliminate forced marriages and genital mutilation (5.3); value unpaid care and promote shared domestic responsibilities (5.4); ensure full participation in leadership and decision-making (5.5); universal access to reproductive health and rights (5.6); equal rights to economic resources, property ownership, and financial services (5.A); promote empowerment of women through technology (5.B), and adopt and strengthen policies and enforceable legislation for gender equality (5.C) (UN General Assembly, 2015). The central global objective is whether these goals will be achieved in all signatory countries by 2030.

Each of these targets has one or more indicators, which provide specific data to track, as noted in Figure 1.3.

Geography, gender, health, and sustainability

It has been well established that improving gender equality generally enables the achievement of better health for all (Shannon et al., 2019; Franz & Ghebreyesus, 2019). Invariably, gender issues are inherently geographical as they happen in time and space. For instance, geographical context is likely to determine to what extent gender equality (SDG 5) can be promoted, in order to support health and well-being. In the context of social determinants of health, reinforcing the intersectionality and interactions between various determinants will depend on where these take place, and the level of equality in particular places. Moreover, one of the challenges in examining gender matters globally is the frequent lack of relevant geographical data that can be used to tease out existing gender inequalities. All too often, global data lump all the indicators together without specific gender indices. Yet Liverman (2018a) argues that besides the lack of disaggregation of

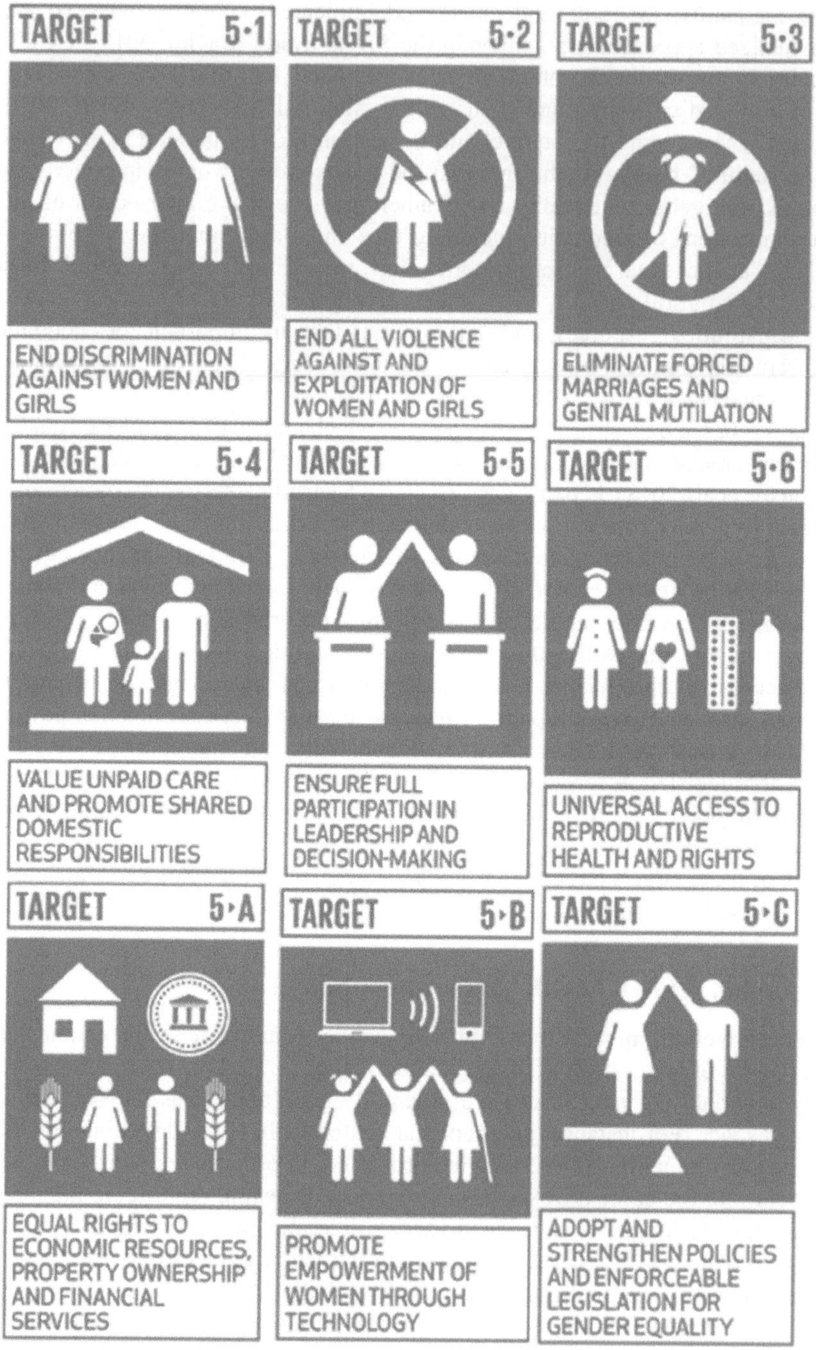

Figure 1.2 Goal 5 Targets

Source: Open Development Mekong (Accessed 5 February 2021)

Target	Indicator(s)
5.1	**5.1.1** Whether or not legal frameworks are in place to promote, enforce, and monitor equality and non-discrimination on the basis of sex
5.2	**5.2.1** Proportion of ever-partnered women and girls aged 15 years and older subjected to physical, sexual, or psychological violence by a current or former intimate partner in the previous 12 months, by form of violence and by age **5.2.2** Proportion of women and girls aged 15 years and older subjected to sexual violence by persons other than an intimate partner in the previous 12 months, by age and place of occurrence
5.3	**5.3.1** Proportion of women aged 20–24 years who were married or in a union before age 15 and before age 18 **5.3.2** Proportion of girls and women aged 15–49 years who have undergone female genital mutilation/cutting, by age
5.4	**5.4.1** Proportion of time spent on unpaid domestic and care work, by sex, age and location
5.5	**5.5.1** Proportion of seats held by women in national parliaments and local governments **5.5.2** Proportion of women in managerial positions
5.6	**5.6.1** Proportion of women aged 15–49 years who make their own informed decisions regarding sexual relations, contraceptive use and reproductive health care **5.6.2** Number of countries with laws and regulations that guarantee women aged 15–49 years access to sexual and reproductive health care, information and education
5.a	**5.a.1** (a) Proportion of total agricultural population with ownership or secure rights over agricultural land, by sex; and (b) share of women among owners or rights-bearers of agricultural land, by type of tenure **5.a.2** Proportion of countries where the legal framework (including customary law) guarantees women's equal rights to land ownership and/or control
5.b	**5.b.1** Proportion of individuals who own a mobile telephone, by sex
5.c	**5.c.1** Proportion of countries with systems to track and make public allocations for gender equality and women's empowerment

Figure 1.3 Targets and Indicators

Source: United Nations. Department of Economic and Social Affairs (Accessed 5 February 2021) © UNITED NATIONS

Copyright: https://sdgs.un.org/goals/goal5

data by gender, the most pervasive distortion associated with aggregation is the lack of attention to within country variations. For instance, Liverman (2018a) presents an example over the MDG period using data from the Ghana Statistical Services where the national indicators showed overall poverty had decreased from 49.4% in 1993 to 22% in 2012. Meanwhile, local level poverty varied from 5.6% in Accra to more than 70% in the Upper West region of Ghana, with even greater variation at the district level (Ghana Statistical Service, 2014). These within nation variations demand critical geographical analysis and interpretation of gender inequities using geographers' unique interdisciplinary skills, in both theory and method, and geographers' engagement with public policy (Sultana, 2018). Health-related phenomenon is understood through a give-and-take relationship between people and their settings, and thus an analysis of gender and health must integrate both contextual and compositional factors. The discipline of geography differentiates between *space* as a physical setting, and *place* as the peopling of that space. Geographical conceptualization of *place* has evolved, advancing our understanding of the constellations of dynamic human interactions (of both men and women), informed by social norms at a wide range of geographical scales. For instance, from a gender lens, access to goods and services can be understood through the complexity of social power relations and the cultural meaning affixed to these power relations (Cummins et al., 2007; Dixon et al., 2014). Hence, geographers have and continue to engage with the SDGs and various global development institutions, through drawing upon long-standing geographical conceptions of development and place-based sociopolitical understandings (Liverman, 2018b). Geographers can thus better inform the progress (or lack thereof) being made with regard to the SDGs across a variety of scales, from the local to the global.

Health geographers have been involved in the application of geographical perspectives, information, and methods to the study of health, disease, and health care (Kitchin & Thrift, 2009). Therefore, health inequalities, and the gendered nature of access to health enabling resources, have been traditional concerns of health geographers (Cummins et al., 2007; Luginaah, 2009; Bisung et al., 2018). Most notably, discussions of inequality and health have become a salient element in gender studies (Bartley, 2016). For instance, in developing countries, there continues to be historically embedded unequal global relationships, composed of unique cultural identities and histories, which inform the special circumstances that define, establish, and maintain health, as it relates to men and women. Consequently, health and other human geographers can provide a spatial understanding of gender (in)equality issues while also generating understanding and insight into the influence of a range of other axis of diversity influencing sex, gender and health, quality of life, and well-being. Geography can offer a unique lens to the study of sex, gender, health, and sustainability by focusing not only on the distributive features of disease and disease services (traditionally conceptualized as space) but also on broadening the discussion of the role of place (Brown & Moon, 2012). To date, conversations about sex and gender within the sub-disciplines of health geography and development geography have been limited and do not reflect the importance of the SDGs.

Clearly, the question of how indicators are constructed and how progress is measured is all important. For example, geographical-based data on various indicators show that, on average, there have been global achievements, and women are outpacing men in tertiary education (Sultana, 2018). Yet this is not translating into equality in various other determinants of health. Consequently, glaring gender gaps remain and the first challenge to confront geographers working on gender and sustainability is the striking lack of geographical data (Liverman,

2018a). There are constant differences in outcomes when country comparisons are made. Gendered health problems in various contexts differ because of the varied environmental, cultural, structural, and economic vulnerabilities of the people they impact. An area that needs attention is the intersection of sex and gender, health and sustainability. By examining why sex or gender matters globally, we rely on interdisciplinary synergies to ensure that a geographical perspective is evident in global sex and gender health research. Geographers do this by establishing an understanding of the geographical variations of sex, gender, health, and sustainability. The SDGs rely on countries to monitor and implement programs to meet the goals, but for many, their ability to achieve them is limited because their own environment, economy, and even social conditions are heavily influenced by processes and policies outside of their borders (Liverman, 2018a). Geographers have the potential to help in the development of indicators, methods, and strategies for assessing and supporting progress toward the SDGs. Geographers can offer a nuanced understanding of scale, statistics, and spatial analysis and our ability to connect social and environmental conditions. Informed by social theory, work aimed at advancing SDGs 5 and others need to pay attention to nature, labor, gender, multiple identities, and bodies (Liverman, 2018b). This book highlights the need for a global perspective on sex and gender matters. We need to continuously monitor health inequalities, collecting data that reflects women's empowerment, income, gender, age, race, ethnicity, migratory status, disabilities, and where people live. This book points to the need for a geographical perspective and sets out the centrality of a geographical contribution to why sex gender matters for the achievement of the SGDs.

Overview of sections

Each of the four sections highlights SDG Goal 5 and its various targets, in addition to the synergies with many other SDGs, and thereby provides a range of contributions across the globe. In addition to highlighting a range of Sub-Saharan African countries, Cambodia, Japan, and the three countries making up North America (Canada, USA, and Mexico) are featured.

Part 1: SDG 5: gender equality and empowerment of women and girls

Chapter 2 employs secondary data to assess the potential for achieving SDGs 3 and 4 on health and gender equity for young people in Ghana. While focusing on SDG 3, the chapter argues that there is a strong interlinkage between SDGs 3 and 4 given the gender inequity that is bias against young females.

Liberia is the focus of Chapter 3, where post-conflict recovery reflects the challenges of gender equality in a dynamic political environment. The authors examine the interpretive politics of gender equality, as articulated by Liberian women on the basis of their experiences as they navigate political, local, and international paradigmatic systems that simultaneously hinder and present opportunities for gender equality.

Japan is highlighted in Chapter 4, where gender statistics at the local level in the city of Mito are discussed. Here, the local government has made the collection of gender statistics inform evidence-based policy-making, and ultimately progress in reaching *SDG #5 Gender Equality*.

A global overview of the patterns in the employment of women in the renewable energy sector is provided in Chapter 5, addressing both SDG 5 (achieve gender equality and empower all women and girls) and SDG 7 (ensure access to affordable, reliable, sustainable, and modern energy for all), and the many synergies between the two goals. The authors discuss global trends, opportunities, and constraints for women's employment in RE, providing estimates about the total share of women in the global RE workforce, and, more specifically, in managerial and leadership positions. Survey findings also shed light upon barriers and opportunities for women's entry, retention, and advancement in the RE sector in various world regional contexts.

Part 2: Target 5.4: value unpaid care and promote shared domestic responsibilities

Chapter 6 provides a contribution to knowledge, information, and geospatial analysis of gender research, aimed at the 2030 SDGs. Employing a methodological approach that applies systems theory, interdisciplinary modeling, and communication, policy conversations that link gender and territorial perspectives are proposed.

A qualitative approach is employed in Chapter 7 to shed light on the lived experiences of immigrant carer-employees in Canada and specifically addresses *SDG 5.4.1: Unpaid Care and Domestic Work*. This research highlights the spatial and temporal tensions of the provision of unpaid eldercare and domestic work in a multi-generational household, for those who are also conducting paid work in the micro-environment of the home. This chapter illustrates that the unequal division of labor persists in many developed nations, and particularly wealthy countries that are primarily made up of immigrants.

The third chapter in this section (Chapter 8) provides an in-depth discussion of resource insecurity and gendered inequalities in health. This is accomplished via an extensive discussion of resource insecurity as reflection of food, water, and energy insecurity at the household and individual levels, as well as how the co-occurrence of these insecurities further worsens the health and economic consequences experienced by women.

The interconnectivity of and synergistic effects between SDGs 5 and 6 are elaborated on in Chapter 9, where universal and equitable access to water, sanitation, and hygiene (WASH) is recognized as fundamental for empowering women and girls in rural Ghana and elsewhere. That is, gender equality (SDG 5) is proposed to be a prerequisite to the overall success of the SDG 6 agenda. The authors suggest that investing in gender equality – where there is equal participation of men and women – can deliver high returns in terms of universal access to safe drinking water.

Chapter 10 is set in the USA, where farmers' markets are the setting for research addressing SDGs Goal 2: End Poverty, and Goal 5: Gender Equality, given that farmer's markets combat food insecurity by providing sustainable, affordable, and accessible food source to women and families.

Part 3: Target 5.6: universal access to reproductive health and rights

Adequate use of antenatal care (ANC) is considered helpful in preventing maternal and child mortality and morbidity, as discussed in Chapter 9 within the context of Ghana. Addressing SDGs 3, 5, and 10, the authors of Chapter 11 highlight the role of internal migration on ANC utilization.

Contributions in Chapter 12 highlight trends in women's empowerment in Sub-Saharan Africa, where results show that women's empowerment, over the long run, is an important preventive strategy for reducing maternal mortality, especially in Sub-Sahara Africa.

The third chapter in this section (Chapter 13) employs a mixed-methods approach to examine the mental health, quality of life, and life experiences of Ghanaian women living with breast cancer. Connections are made to SDG 1, 3, 5, and 8, and specifically 5.A, 8.1, and 8.2. Further, the results make clear that access and utilization of health care is a clear impediment for many breast cancer patients, and especially those in rural parts of the country.

Chapter 14 provides an overview of the determinants of breastfeeding in Cambodia, noting that breastfeeding is an integral element in achieving the SDGs, as it is directly linked with Goal 2 – end hunger and improved nutrition, Goal 3 – ensure healthy lives and promote well-being, and Goal 12 – ensure sustainable consumption and production.

The concluding chapter (Chapter 15) provides a thematic overview of the volume, demonstrates the interconnections between gender (in)equality and sustainable development, and lays out future agenda to advance sustainable gender geography.

Notes

1 The World Health Organization defines female genital mutilation as "all procedures involving partial or total removal of the external female genitalia or other injury to the female organs whether for cultural or other non-therapeutic reasons."

2 Gender parity is defined as having a gender parity index [GPI] value between 0.97 and 1.03.
Notes: Gender parity index on adjusted net enrollment rate is used as primary source, and where administrative data is not available, household survey data is used.

References

Agarwal, B. (2018). Gender equality, food security and the sustainable development goals. *Current Opinion in Environmental Sustainability*, *34*, 26–32.

Ameyaw, E. K., Tetteh, J. K., Armah-Ansah, E. K., Aduo-Adjei, K., & Sena-Iddrisu, A. (2020). Female genital mutilation/cutting in Sierra Leone: Are educated women intending to circumcise their daughters? *BMC International Health and Human Rights, 20*(1), 1–11.

Anand, P., Hunter, G., Carter, I., Dowding, K., Guala, F., & Van Hees, M. (2009). The development of capability indicators. *Journal of Human Development and Capabilities, 10*(1), 125–152.

Azcona, G., & Bhatt, A. (2020). Inequality, gender, and sustainable development: Measuring feminist progress. *Gender & Development, 28*(2), 337–355.

Bartley, M. (2016). *Health inequality: An introduction to concepts, theories and methods.* London, UK: John Wiley & Sons.

Bisung, E., Dixon, J., & Luginaah, I. (2018). The past, present and future contributions of health geography. In *Routledge handbook of health geography*. London, UK: Taylor & Francis.

Boateng, G. O., Kuuire, V. Z., Ung, M., Amoyaw, J. A., Armah, F. A., & Luginaah, I. (2014). Women's empowerment in the context of millennium development goal 3: A case study of married women in Ghana. *Social Indicators Research, 115*(1), 137–158.

Brown, C. S., Ravallion, M., & Van De Walle, D. (2017). *Are poor individuals mainly found in poor households? Evidence using nutrition data for Africa* (No. w24047). National Bureau of Economic Research. Cambridge, MA, USA.

Brown, T. I. M., & Moon, G. (2012). Geography and global health. *The Geographical Journal, 178*(1), 13–17.

Butler, J. (2011). *Gender trouble: Feminism and the subversion of identity*. London, UK: Routledge.

Carr, E. (2005). Development and the household: Missing the point? *GeoJournal, 62*, 71–83.

Choe, S. A., Cho, S. I., & Kim, H. (2017). Gender gap matters in maternal mortality in low and lower-middle-income countries: A study of the global Gender Gap Index. *Global Public Health, 12*(9), 1065–1076.

Cummins, S., Curtis, S., Diez-Roux, A. V., & Macintyre, S. (2007). Understanding and representing "place" in health research: A relational approach. *Social Science & Medicine, 65*(9), 1825–1838.

de Leon, R. G. P., Ewerling, F., Serruya, S. J., Silveira, M. F., Sanhueza, A., Moazzam, A., Becerra-Posada, F., Coll, C. V., Hellwig, F., Victora, C. G., & Barros, A. J. (2019). Contraceptive use in Latin America and the Caribbean with a focus on long-acting reversible contraceptives: Prevalence and inequalities in 23 countries. *The Lancet Global Health, 7*(2), e227–e235.

Dixon, J., Tenkorang, E. Y., & Luginaah, I. (2011). Ghana's national health insurance scheme: Helping the poor or leaving them behind? *Environment and Planning C, 29*(6), 1102–1115.

Dixon, J., Luginaah, I., & Mkandawire, P. (2014). The national health insurance scheme in Ghana's Upper West region: A gendered perspective of insurance acquisition in a resource-poor setting. *Social Science and Medicine, 122*, 103–112.

Doyal, L. (2000). Gender equity in health: Debates and dilemmas. *Social Science & Medicine, 51*(6), 931–939.

Esquivel, V., & Sweetman, C. (2016). Gender and the sustainable development goals. *Gender & Development, 24*(1), 1–8.

Franz, C., & Ghebreyesus, T. A. (2019). *The road to universal health coverage: Innovation, equity, and the new health economy*. Baltimore, MD, USA: JHU Press.

Kim, E. M. (2017). Gender and the sustainable development goals. *Global Social Policy, 17*(2), 239–244.

Kim, S., & Seo, J. (2017). A study on the influence of gender gap on economic structural improvement and economic growth. *Journal of the Society of Disaster Information, 13*(4), 499–510.

Kitchin, R., & Thrift, N. (2009). *International encyclopedia of human geography*. Amsterdam, The Netherlands: Elsevier.

Liverman, D. M. (2018a). Geographic perspectives on development goals: Constructive engagements and critical perspectives on the MDGs and the SDGs. *Dialogues in Human Geography, 8*(2), 168–185.

Liverman, D. M. (2018b). Development goals and geography: An update and response. *Dialogues in Human Geography, 8*(2), 206–211.

Luginaah, I. (2009). Health geography in Canada: Where are we headed? *The Canadian Geographer, 53*(1), 91–99.

Meinzen-Dick, R., Kovarik, C., & Quisumbing, A. R. (2014). Gender and sustainability. *Annual Review of Environment and Resources, 39*, 29–55.

Montesanti, S. R. (2015). The role of structural and interpersonal violence in the lives of women: A conceptual shift in prevention of gender-based violence. BMC Women's Health, 15, 19 DOI 10.1186/s12905-015-0247-5

Moss, N. E. (2002). Gender equity and socioeconomic inequality: A framework for the patterning of women's health. *Social Science & Medicine, 54*(5), 649–661.

Nussbaum, M. C. (2008). Creating capabilities: The human development approach and its implementation. *Hypatia, 24*(3), 211–215.

OXFAM (2017). Governance in Tajikistan: Evaluation of the women smallholder farmer advocacy campaign. *Effectiveness Review Series 2015–16*. Available from: https://oxfamilibrary.openrepository.com/bitstream/handle/10546/620286/er-governance-tajikistan-effectiveness-review-280617-en.pdf;jsessionid=0013C8762F166DC0C0D5264EBE8FE272?sequence=1

Phillips, S. P. (2005). Defining and measuring gender: A social determinant of health whose time has come. *International Journal for Equity in Health, 4*(1), 1–4.

Presiden Republik Indonesia (2004). Available from: www.wcwonline.org/pdf/lawcompilation/Indonesia-Regarding-Elimination-of-Violence-in-Household.pdf

Pyles, L. (2008). The capabilities approach and violence against women: Implications for social development. *International Social Work, 51*(1), 25–36.

Ridgeway, C. L. (2011). *Framed by gender: How gender inequality persists in the modern world*. London, UK: Oxford University Press.

Risman, B. J., Lorber, J., & Sherwood, J. H. (2012, August). Toward a world beyond gender: A utopian vision. In *American Sociological Association Annual Meeting*. www.ssc.wisc.edu/~wright/ASA/Risman-Lorber-Sherwood% 20Real (Vol. 20).

Ritchie, H., Roser, M., Mispy, J., & Ortiz-Ospina, E. (2018). Measuring progress towards the sustainable development goals. *SDG-Tracker. org, website*. Available from: https://sdg-tracker.org/gender-equality

Rocheleau, D. (1995). Maps, numbers, text, and context: Mixing methods in feminist political ecology. *The Professional Geographer, 47*(4), 458–466.

Sen, A. K. (1982). *Poverty and famines: An essay on entitlement and deprivation*. London, UK: Oxford University Press.

Sen, A. K. (1990). Gender and cooperative conflicts. In I. Tinker (Ed.), *Persistent inequalities: Women and world development*. New York: Oxford University Press.

Shannon, G., Jansen, M., Williams, K., Cáceres, C., Motta, A., Odhiambo, A., Eleveld, A., & Mannell, J. (2019). Gender equality in science, medicine, and global health: Where are we at and why does it matter? *The Lancet, 393*(10171), 560–569.

Sharma, R. R., Chawla, S., & Karam, C. M. (2021). Global gender gap index: world economic forum perspective. In *Handbook on diversity and inclusion indices*. Edited by: Eddy S. Ng, Christina L. Stamper, Alain Klarsfeld, Yu (Jade) Han. Northampton, MA, USA: Edward Elgar Publishing.

Song, J., & Kim, E. M. (2013). A critical review of gender in South Korea's official development assistance. *Asian Journal of Women's Studies, 19*(3), 72–96.

Sultana, F. (2011). Water, culture, and gender: An analysis from Bangladesh. In *Water, cultural diversity, and global environmental change* (pp. 237–252). Dordrecht: Springer.

Sultana, F. (2018). An (other) geographical critique of development and SDGs. *Dialogues in Human Geography, 8*(2), 186–190.

Tolhurst, R., Amekudzi, Y. P., Nyonator, F. K., Bertel Squire, S., & Theobald, S. (2008). "He will ask why the child gets sick so often": The gendered dynamics of intra-household bargaining over healthcare for children with fever in the Volta Region of Ghana. *Social Science & Medicine, 66*(5), 1106–1117.

UN (2015). *The investment case for education and equity*. New York: UNICEF.

United Nations Development Programme (UNDP). (2003). *Human development report, 2003. The millennium development goals: A compact among nations to end human poverty*. New York, NY: Oxford University Press.

United Nations Development Programme (UNDP). (2015). Millennium Development Goals Report 2015. Available from: https://www.un.org/millenniumgoals/2015_MDG_Report/pdf/MDG%202015%20rev%20(July%201).pdf

UNESCO Institute for Statistics global databases (2019, September). *Gender and education*. Available from: https://data.unicef.org/topic/gender/gender-disparities-in-education/

UN General Assembly (2015). Transforming our world: The 2030 Agenda for Sustainable Development, 21 October 2015, A/RES/70/1. Available from: www.refworld.org/docid/57b6e3e44.html [accessed 28 February 2021]

UN Women (2018). Turning promises into action. Gender equality in the 2030. Available from: https://www.unwomen.org/-/media/headquarters/attachments/sections/library/publications/2018/sdg-report-gender-equality-in-the-2030-agenda-for-sustainable-development-2018-en.pdf?la=en&vs=4332

UN Women (United Nations Entity for Gender Equality and the Empowerment of Women) (2019). *Progress of the world's women 2019–2020: Families in a changing world*. New York: UN Women.

Vyas, S., & Jansen, H. A. (2018). Unequal power relations and partner violence against women in Tanzania: A cross-sectional analysis. *BMC Women's Health, 18*(1), 1–12.

World Economic Forum (2020). *The global gender gap report 2020: Insight report*. Cologny, Switzerland: World Economic Forum.

World Health Organization (2017). *Country support package for equity, gender and human rights in leaving no one behind in the path of universal health coverage* (No. WHO/FWC/GER/17.1). Geneva: World Health Organization.

Part 1

SDG 5

Gender equality and empowerment of women and girls

This first section of the collection provides an overview of *Goal 5*, while concentrating on several of the *9 Targets* (see Introductory Chapter 1, Figure 2). In so doing, the four chapters in this section provide a broad look at the importance of gender equality and empowerment of women and girls for overall progress in sustainable development and improved quality of life and health outcomes for all. Further, this section illustrates the interconnectivity between all 17 SDGs, and specifically *Goal 3: Good Health and Well-Being, 4: Quality Education*, and *7: Affordable and Clean Energy*, while highlighting countries in sub-Saharan Africa (Ghana, Liberia) and Japan.

DOI: 10.4324/9780367743918-2

2 Gender, adolescents, and achieving the Sustainable Development Goals in Ghana

Rev. Adobea Yaa Owusu

Introduction

Adolescents/young people are critically important to society. Their sheer numbers, energy, and strength are important resources. They also constitute the future generation that will take over the helm of affairs and prepare the next generation for the continuity of society. Healthwise, adolescents/young people are particularly important as they are used to gauge other critical aspects of the health status of nations. For instance, people aged 15–24 are used as a proxy measure for new HIV infections (incidence) internationally. Currently, the world has the largest generation of young people in history: 1.8 billion people between the ages of 10 and 24 years (WHO, 2018). This cohort represents a powerhouse of human potential that could transform health and sustainable development. Their engagement is critical to achieving the 2030 Agenda for Sustainable Development (WHO, 2018). Yet in Ghana, there is widespread lack of prime socio-economic opportunities for adolescents/young people, particularly females, to achieve their optimum potential. This has implications for the potential of Ghana to achieve the United Nations' Sustainable Development Goals (SDGs).

Concern for adolescents' health and well-being is acknowledged both internationally and within Ghana. A clear case for investing in the health and well-being of young people has been made by WHO (2017a, 2017b, 2018) and others (Sheehan et al., 2017) over the last decade. Without doubt, the health and well-being of the planet's largest generation of adolescents will shape both the future of the world's health and the achievement of the SDGs related to health (SDG3), education (SDG4), gender equality (SDG5), and hence help in reducing inequalities (SDG10) (Hashiguchi, 2016). The emergence of the SDGs led to the emergence of a renewed and expanded focus on adolescent health and well-being in Ghana.

There is paucity of literature regarding adolescents in relation to the SDGs. Most of the available literature on the topic is limited to education (Chongcharoentanawat et al., 2016). This is applicable to Ghana (Konadu and Osei, 2020), where there is also limited data, particularly indicator-based data, on the SDGs including gendered aspects of the data. Furthermore, there is paucity of studies that focus on adolescent health in Ghana (Dako-Gyeke et al., 2020). The SDGs, which followed the MDGs, have arguably come to the rescue of adolescent

DOI: 10.4324/9780367743918-3

women in Ghana and give them another chance to attain optimum development. Primarily, this chapter reviews SDG3 within the context of Ghana, and guesstimates its potential for improving the health of female adolescents in Ghana by 2030, based on select goals and indicators. The chapter focuses on reproductive health, including family planning (FP) and HIV/AIDS. Primarily, it positions female adolescents as a particularly vulnerable group, mostly informed by their youthful age and gender, and inherent inequities in gender opportunities in Ghana. The latter is further informed by culture and general dearth of socio-economic advantage. Relying on secondary data, the chapter also illuminates the SDG on gender equality without which the human progress touted by the SDGs would be in vain (United Nation, no date).

For context, the chapter reviews the documented end-stage milestones/achievements of the MDGs in Ghana with respect to gender and health well-being, generally and for adolescent females in particular. This forms a reference point for the SDGs, particularly SDG3 and SDG5. With a focus on SDG3 (Good Health), the chapter reviews the current issues in respect of the health and health-care seeking tendencies of adolescent girls and young women in Ghana. Given the background of underdevelopment and relatively entrenched cultural dictates, particularly regarding the comparative socio-economic disadvantage of females, an important thematic context of the chapter is the gender vulnerability hypothesis.

National context

Ghana is located at West African's Gulf of Guinea, between latitude 7.9465° N and longitude 1.0232° W (Figure 2.1). It covers an area of approximately 238,540 km^2. It is ranked 24th by size among the 54 African countries (FAO, 2005). It is bordered on the north by Burkina Faso, on the east by Togo, and on the west by Cote d'Ivoire. The north–south expanse of the country is about 670 km, and a maximum of 560 km for the east–west (FAO, 2005). It is only few degrees north of the equator and is on the Greenwich meridian line which passes through the seaport of Tema, about 24 km to the east of Africa. The country is currently divided into 16 administrative regions (from ten regions) since January 2019. The national capital, Accra, is located at the mid-eastern part of Ghana's southern coastline (Figure 2.2).

Ghana's last population census in 2010 recorded a total population of 24,658,823 (GSS, 2012). Currently (March 2020), there is an estimated population of 30.9 million, with a growth rate of 2.15% (World Population Review, 2020). The census revealed a youth bulge unprecedented in Ghana's history; the largest group to be entering into adulthood was young people aged 10–24 years (GSS, 2012). This cohort of young people constituted 31.8% of the total population (GSS, 2012). Approximately 61.5% of the total population of Ghana was below age 25 years. A little less than a quarter of the population were adolescents aged 10–19 years. This was made up of about 12% and 11%, respectively, of 10–14- and 15–19-year-olds. The proportion of the male adolescents in Ghana is higher than that of female adolescents. The urban–rural variation shows 24.4% male adolescents (aged 10–19

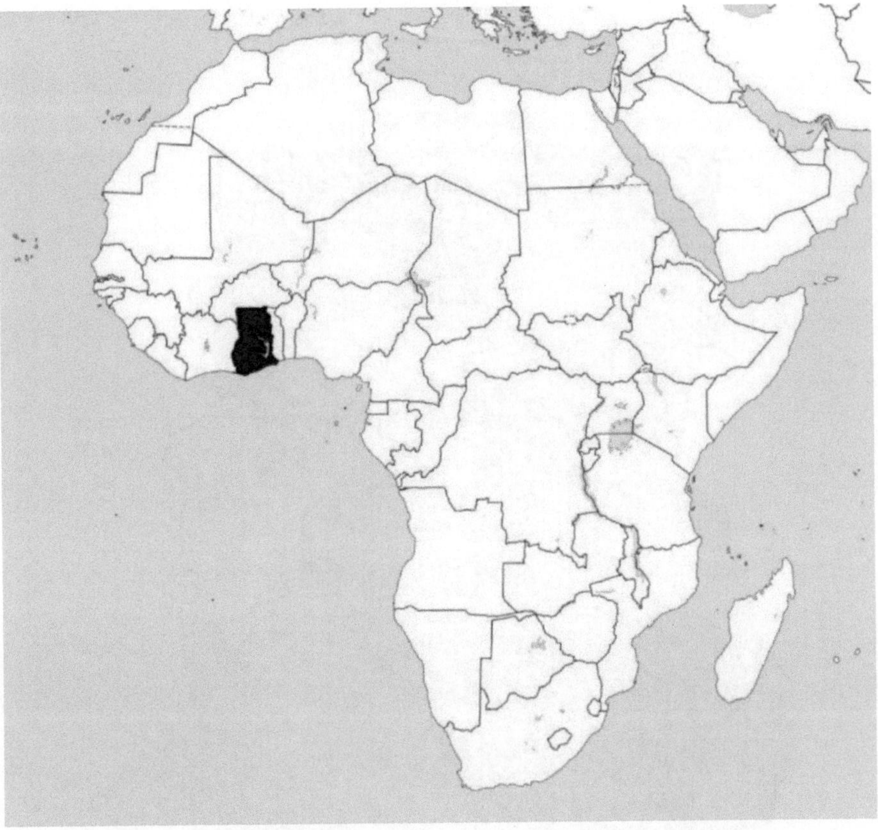

Figure 2.1 Map of Africa showing Ghana

Source: https://maps-ghana.com/ghana-africa-map. (Accessed October 15, 2020) © 2020 Copyright: Newebcreations

years) in the rural areas, compared to 21.4% female adolescents. Adolescents in urban areas are 22% for both males and females (GSS, 2012).

The very high proportion of Ghanaian youth implies a high dependency ratio. That is, a few adults support young and, even, older people too. This also has a negative repercussion on the demographic dividend. Demographic dividend implies the socio-economic opportunities emanating from having higher proportions of the working-age group as a result of lowering birth rates (Lutz et al., 2019). These two demographic scenarios in Ghana will potentially militate against the country's ability to achieve its SDG-related objectives. Specifically, failure to achieve the demographic dividend will influence the number of youths the country has to grapple with in achieving its SDGs. Rwanda, for instance, is mitigating this impediment to achieving its SDGs by striving toward a dividend (Republic of Rwanda, 2019).

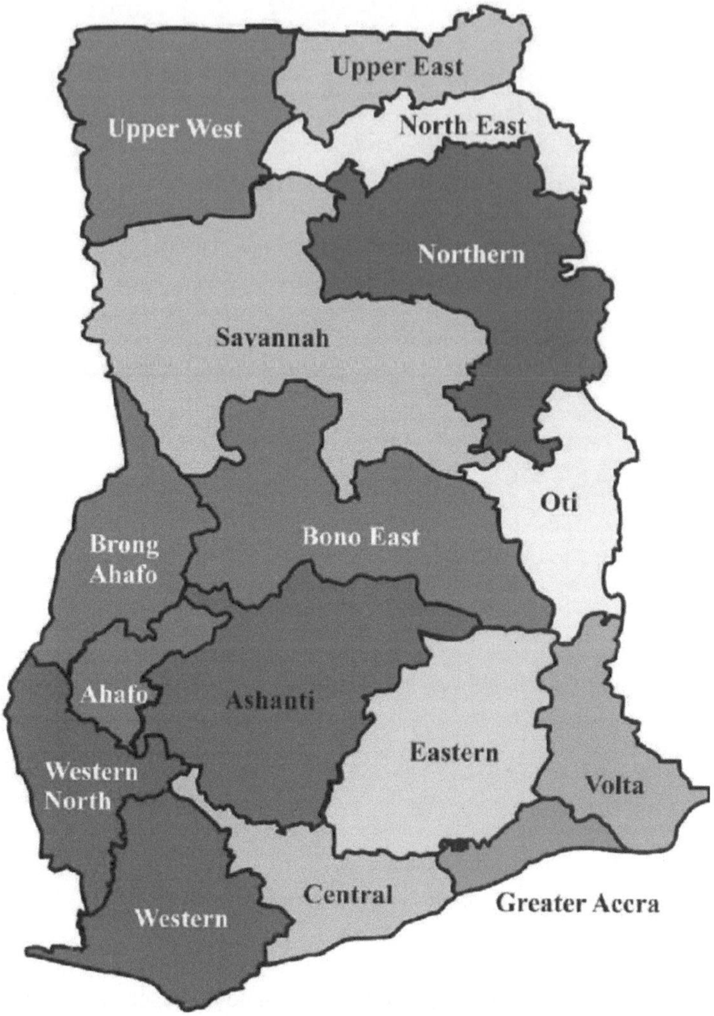

Figure 2.2 Map of Ghana showing its 16 administrative regions

Source: The Permanent Mission of Ghana to the United Nations. Available at: www.ghanamissionun.org/wp-content/uploads/2020/09/Ghana_Regional_Map.png. (Accessed 25 March 2021)

Importantly, the health well-being of young people is shaped by the socio-cultural and economic contexts in which they live. In Ghana, the youths are faced with multiple and interrelated social, economic, and health challenges. Gender inequality, sexual coercion, early marriage, high levels of teenage pregnancy, unsafe abortion, sexually transmitted infections (STIs), including HIV/AIDS, are among the sexual and reproductive health problems faced by many young

Ghanaians (GSS, 2012). These are further complicated by limited access to reproductive health information and good-quality adolescent and youth-friendly reproductive health services (Grindlay et al., 2018). Currently, there is a concurrent lack of prime socio-economic opportunities for young people between the ages of 10 and 24 years despite them having the highest educational opportunities ever. Particularly, there is a very high level of unemployment among young people in Ghana (Amissah and Nyarko, 2017; Domfe and Oduro, 2018) contributing to their documented poor mental health status and its devastating psychological dimensions (Amissah and Nyarko, 2017). Ghana's youth unemployment is said to be among the highest in the world (Amissah and Nyarko, 2017), ranging up to 30.8% in the national capital (Amankrah, 2003). Suicide among teenagers and young people in Ghana is rising, resulting partly from the lack of socio-economic opportunities (Nii-Boye Quarshie et al., 2015) although suicide is not limited to young people alone. Against this background, the extended family support system in Ghana is waning (Owusu, 2007; Lund and Agyei-Mensah, 2008).

Definitions of adolescence and youth

An adolescent is a person within the age group of 10–19 years. Traditionally, the term young person connotes those between 10 and 24 years (WHO, 2014). The United Nations (1981) defines the youth to include all people aged 15–24 years. These definitions provide a universal consensus on the biological delineation of the youth. However, national definitions of the designated age bracket for the youth vary due to differences in national policies. In Ghana, for instance, the National Youth Policy, informed by the African Youth Charter (African Union, 2006) classifies all people 15–35 years as youth (Republic of Ghana, 2004). Thus, in Ghana, the classification of youth overlaps adolescents and children between 15 and 19 years and beyond the 24-year-old limit used by the United Nations (GSS, 2012). Ghana's definitions and classification of the three sub-population groups (children, adolescents, and youths) overlap. The term "adolescent" is used synonymously with "teenager." The latter ranges from 13 to 19 years. The terms early adolescence (10–14), late adolescence (15–19), and post adolescence (20–24) have also been used for purposes of clarity and ease of understanding.

MDGS and the SDGs in the context of Ghana

The inability of many nations to achieve progress on the MDGs called for a subsequent agenda aimed at sustaining the end-stage progress of the MDGs and advancing world-wide development even further. This led to the implementation of the SDGs in 2015, also known as the Agenda 2030, as a sequel to the MDGs. The succession of the MDGs by the SDGs can be appropriately described as a timely intervention for the rescue of Ghanaian adolescent/young women in terms of attainment of their optimum development. The MDGs and SDGs have been cited as the main criteria for determining and judging progress in international development efforts (Karver, Kenny and Sumner, 2012; Blaboe, 2019). It is, therefore, imperative to assess the developmental progress of Ghana in the light of the philosophies of the MDGs and SDGs.

On the basis of their national relevance, Ghana adopted 17 targets and 36 indicators out of the 21 targets and 60 indicators globally adopted to monitor the MDGs (NDPC and UN Ghana, 2015). The next section reviews the documented end-stage milestones/achievements of the MDGs in Ghana with respect to gender and broad health well-being, generally and for females in particular, with emphasis on adolescent/young girls.

End-result progress of MDGs in Ghana

Overall, Ghana fairly lagged behind achieving the MDGs on health, gender equality, and women's empowerment (Table 2.1).

The end-stage status of the achievement of the MDGs in Ghana (Table 2.1) largely depicts some gender disadvantage in the health status of young people, particularly young females. Women are directly affected by maternal mortality, particularly younger women in their reproductive years. The same applies to the disastrous consequences of inadequate contraceptive prevalence, antenatal care coverage, and access to/use of skilled assistance at birth. Additionally, Ghana's failed attempt to meet the MDG on HIV has a strong toll on the youth, particularly females, as this chapter shows. These draw attention to the importance of ensuring gender equality in Ghana by 2030, as part of a global commitment to position gender equality as a basic human right and a central ingredient to achieving all other facets of development and human progress internationally, as SDG5 stipulates (United Nation, no date). Gender equality necessitates empowering all women and girls in all facets of society, to the level of men and boys, if not better. Premised on SDG5, gender equality is a two-pronged approach to facilitating sustainable development, both as a specified goal and as a fundamental prerequisite to advancing progress in other areas of development (UN Women, 2018).

Several factors impede access to and use of adequate reproductive health in Ghana. These include poor geographical and financial access to healthcare, including inadequate health insurance coverage, poor healthcare quality, and poor(er) socio-economic status (Ghana Statistical Service (GSS), Ghana Health Service (GHS) and ICF, 2015; GSS, GHS and ICF, 2018). For women in Ghana, patriarchal norms constitute another delimiting factor.

The Sustainable Development Goals

There was much euphoria and enthusiasm surrounding the advent of the SDGs in 2015, but there was little empirical evidence regarding how much would be required of countries to achieve the goals (Konadu and Osei, 2020). The SDGs comprise 17 goals with 169 targets and 230 indicators (Inter-Agency and Expert Group on SDG Indicators [IAEG-SDGs], 2016; Blaboe, 2019). The focus of this chapter is on SDG3 and its select targets and indicators, shown in the next section

Table 2.1 Progress in MDGS Goals 4, 5, and 6 in Ghana as of September 2015

MDG targets	*Indicators*	*General description of end result*	*Specifics of progress made/not made*
Goal 4: Reduce child mortality			
Reduce by two-thirds, between 1990 and 2015, the under-five mortality rate	4.1 Under-5 mortality rate (40 deaths per 1,000 live births)	Not achieved but significant progress	60 deaths per 1,000 live births
	4.2 Infant mortality rate	Achieved	19 deaths per 1,000 live births
	4.3 Proportion of 1-year-old children immunized against measles	Not achieved but significant progress	86% (against the 100% target)
Goal 5: Improve maternal mortality			
a) Reduce by three quarters, between 1990 and 2015, the maternal mortality ratio (190 deaths per 100,000 live births)	a) 5.1 Maternal mortality ratio	Not Achieved	380 deaths per 100, 000 live births
	b) 5.2 Proportion of births attended by skilled health personnel	Not achieved but significant progress	55.3%
b) Achieve, by 2015, universal access to reproductive health	b) 5.1 Contraceptive prevalence rate (CPR)	Not achieved but significant progress	27%
	b) 5.2 Antenatal care coverage (at least one visit)	Not achieved but significant progress	97%
Goal 6: Combat HIV/AIDS, malaria and other diseases.			
a) Have halted by 2015 and begun to reverse the spread of HIV/AIDS	a) 6.1 HIV prevalence among population aged 15–24 years	Not achieved but significant progress	3.0
b) Achieve, by 2010, universal access to treatment for HIV/AIDS for all those who need it	b) 6.1 Proportion of population with advanced HIV infection with access to antiretroviral drugs	Not achieved but significant progress	47.4%
c) Have halted by 2015 and begun to reverse the incidence of malaria and other major diseases	Proportion of children under 5 sleeping under insecticide-treated bed nets	Not achieved but significant progress	18%

Source: NDPC and UN Ghana (2015)

Table 2.2 Goal 3. Ensure healthy lives and promote well-being for all at all ages: Select SDG3 targets and indicators

	Targets	*Indicators*
3.1	By 2030, reduce the global maternal mortality ratio to less than 70 per 100,000 live births	Maternal mortality ratio Proportion of births attended by skilled health personnel
3.2	By 2030, end preventable deaths of newborns and children under 5 years of age, with all countries aiming to reduce neonatal mortality to at least as low as 12 per 1,000 live births and under-5 mortality to at least as low as 25 per 1,000 live births	Under-five mortality rate Neonatal mortality rate
3.3	By 2030, end the epidemics of AIDS, tuberculosis, malaria and neglected tropical diseases and combat hepatitis, water-borne diseases and other communicable diseases	Number of new HIV infections per 1,000 uninfected population by sex, age, and key populations

Source: (IAEG-SDGs, 2016, pp. 4–6).

(Table 2.2). These propose to remedy the multiplicity of challenges that plague the youth, especially female adolescents, in recent times in the areas of health and empowerment. Furthermore, they have direct potential in resolving serious health/healthcare challenges confronting adolescents/young people in Ghana. Gender equality (SDG5) is particularly an integrated approach to achieving all the other SDGs (UN Women, 2018).

The SDGs and adolescent reproductive health in Ghana

The next section focuses on the SDG targets on reproductive health covered in this chapter (Table 2.3). It mostly covers contraception and HIV/AIDS. Over a little more than the last decade, there has been a general improvement in the SDG indicators on maternal and child health in Ghana. However, these have seen almost no progress since the end of the MDGs in 2015 (Table 2.3). For example, the projected estimate for maternal mortality ratio (MMR) for 2018–2021 was 181 per 100,000 live births (Ministry of Health [MOH], 2017). SDG3 target 1 stipulates a reduction of MMR to less than 70 per 100,000 live births by 2030. This means that if the current trend is sustained, Ghana cannot achieve this target.

Contraceptive use and voluntary abortion

In Ghana, although current awareness of FP, particularly modern ones, is 99% (GSS, GHS and ICF, 2018), the FP acceptor rate for 2017 was 36.6%, an increase from 28.7% in 2015 (GHS, 2017, 2018). In 2017, the contraceptive prevalence rate was 25% among women aged 15–49 years, and 20% of these were modern methods. Thirty-one percent of currently married women aged 15–49 used some

Table 2.3 Trends in maternal and child health, Ghana: 2008–2019

Indicator	*Year*										
	2007	*2008*	*2009*	*2010*	*2011*	*2012*	*2013*	*2014*	*2015*	*2016*	*2017*
Maternal mortality ratio[1]	580	451		350			380	144	142	151/163.5/181	310
Proportion of births attended by skilled health personnel[2]	55	59/ 55						74			79.4
Under-five mortality rate[3]	82	80	66	66	82	66	52	60	52	52	52
Neonatal mortality rate[4]	29	30	28	28	32	28	25	29	25	25	25

Sources: GSS (2012); UNDP, Ghana and NDPC/GOG (2012); MOH (2017); MOH and UNDP (2011); GSS, GHS and ICF, 2015, 2018); GHS (2017)

Notes: 1 Per 100,000 live births

2 %

3 Deaths per 1000 live births

4 Deaths per 1000 live births

method of contraception, and 25% used a modern method. Also, 38% and 31% of sexually active unmarried women aged 15–49 years used any method, and a modern method, respectively. Thus, the use of modern methods increased, within the decade, from 14% in 2007 (GSS, GHS and ICF, 2015, 2018). This contraceptive prevalence rate means that there is an unmet need for contraception.

The 2014 Ghana Demographic and Health Survey (GDHS), which is nationally representative, showed 51% and 34% of married women aged 15–19 and 20–24 years, respectively, had an unmet need for FP. Unmet need for FP for women 15–49 years old was 30%, from 36% in 2008 (ISSER, 2017). Six percent and 21% of both married women and sexually active unmarried women 15–19 and 20–24 years old, respectively, were using a modern FP method. GDHS 2014 further showed that 22.8% of women aged 15–49 used any contraceptive method, and 18.2 used any modern method in 2014. For married women, 26.7% used any method, and 22.2% used any modern method. For adolescents aged 15–16 years, 8.7% used any method, and 6.3% used any modern method. In all 28.6% of women aged 20–24 used any method, and 21.1% of them used any modern method. For sexually active unmarried women, 44.5% used any method, and 31.7% used any modern method. In all, 43.7% of sexually active unmarried adolescents aged 15–19 years and 53.4% of those aged 20–24 years used any FP method.

The use of traditional methods has also decreased over time in favor of modern methods. For instance, between 2007 and 2017, the use of modern methods increased by 10.4% for currently married women. Both currently unmarried and married women in urban areas are less likely to use a modern contraceptive method than their rural counterparts. Additional explanatory variables for modern FP use in Ghana are educational levels, wealth quintiles, and region of residence (GSS, GHS and ICF, 2018).

In Ghana, contraception is focused on women both culturally and in terms of healthcare delivery strategy. Yet, several impediments exist against the use of contraception by young people, particularly females. These include poor access, cost, rumors, and myths on purported side effects. There is personal, religious, or partner opposition to FP, and an entrenched belief that it leads to promiscuity in women (Grindlay et al., 2018). Additionally, cultural tenets such as pronatilism and norms that generally stigmatize the use of contraception persist. Adolescent girls and unmarried young women particularly suffer stigmatization for the use of contraception. This discourages them from accessing it. In some cultures, particularly the northern/Islamic strongholds, women are required to seek the permission of their romantic partners, even husbands, before accessing and using contraception. Failure to do this can be a basis for divorce as such women are accused of needing contraceptives due to extramarital affairs/immoral behaviors.

Also, Ghanaian females are culturally groomed and expected to be subservient during coitus. They are not expected to negotiate sex but to respond to a man's sexual advances, albeit without fail, if in a marital or steady relationship (Adinkrah, 2017). Ghanaian women, especially adolescents/younger women, thus have poor skills for negotiating the use of contraception and safe sex (Grindlay et al., 2018). The Ghana Health Service tried to work toward actively involving males in contraception to resolve some of these problems, but the results have been insignificant. These factors may underlie the high rates of teen pregnancy and motherhood. In Ghana in 2017, a national average of 14.4% girls 15–19 years had begun childbearing. Seven out of ten regions had higher rates than the national average, ranging from the highest of 18.9% in the Western Region to 7.0% in the Greater Accra Region. 32.3% of 19-year-olds had begun childbearing, while 12% had had a live birth and 3% were pregnant with their first child. However, teenage pregnancy and child birth is steadily declining. In 1988, 23% of teenage women had begun childbearing, 22% in 1993, 14% in both 1998 and 2003, 15% in 2007, and 14% in 2014 (GSS, GHS and ICF, 2018; Boah, Bordotsiah and Kuurdong, 2019).

Against the background that 31% of pregnancies in Ghana are unwanted or wrongly timed (GSS, GHS and ICF, 2018; Boah, Bordotsiah and Kuurdong, 2019), there is also a very high level of induced abortion, particularly among young women in Ghana. An estimated 15–30% of maternal deaths in Ghana are from abortions (Rominski and Lori, 2014; Boah, Bordotsiah and Kuurdong, 2019). Between 2007 and 2017, induced abortion increased from 5% to 7% for women aged 15–49 years (GSS, GHS and ICF, 2018; Boah, Bordotsiah and Kuurdong, 2019). Of these, 21% mentioned not being ready for the baby as the main reason for the abortion. For women aged 15–19, 3.3% had induced abortion within the past five years, 15.9% for those aged 20–24, and 24.1% for 25–29-year-olds. 19.6% had ever had an induced abortion (GSS, GHS and ICF, 2018, p. 95).

More importantly, often, such high levels of abortion are done through unsafe, life-threatening means, and by quacks. 27% of women aged 15–49 years who had had induced abortion within the past five years prior to 2017 used non-medical methods (GSS, GHS and ICF, 2018, p. 95). This has led to the deaths of a high number of young women through unsafe abortion and sepsis, leaving some with permanent consequences including lifetime infertility. Abortion is thus an important contributory factor to the high maternal morbidity and mortality in Ghana (GSS, GHS and ICF, 2018; Boah, Bordotsiah and Kuurdong, 2019). Such high levels of abortion, particularly, unsafe abortion, underscore unmet reproductive health needs for young women.

In Ghana, despite a partial legalization of induced abortion in 1985 (Ahiadeke, 2001), limited access to legal abortion services, socio-cultural barriers, cost, and high stigmatization of abortion limit the use of safe abortion services (Payne et al., 2013; Boah, Bordotsiah and Kuurdong, 2019). The high stigma associated with induced abortion and the often clandestine manner in which it is done in Ghana make it difficult to accurately capture its extent. Women in their reproductive ages also lack adequate information about abortion services. For instance, in 2007 and 2017, only 4% and 11%, respectively, of a nationally representative sample of women aged 15–49 years knew that abortion was partially legal in Ghana. Furthermore, only 38.4% of the 2017 sample could correctly state the fertile period during their ovulatory cycle. This increased slightly from 36.3% in 2014. Delays in policy formulation for comprehensive abortion services, which take cognizance of the partial legalization, have also been another limitation (GHS, 2018).

HIV/AIDS in Ghana

Ghana is experiencing a generalized HIV epidemic Ghana AIDS Commission (GAC) (GAC, 2019). This means a prevalence of more than 1% in the general population (GHS and NACP, 2019; Owusu, 2020). The incidence data show a sustained decline in the number of new infections over the past 12 years in Ghana, from a high of 0.14% in 2008 to 0.10% in 2019 and an estimated 0.04% by 2025. The age group spanning 20–34 years accounts for most new infections (Table 2.4). Overall, new HIV infections are estimated to decline from 21,059 in 2015 to 19,931 in 2018, and eventually decline consistently to 8,752 in 2025 (GHS and NACP, 2019). The HIV prevalence rate among the 15–19 and 20–24 age groups has declined over the last decade (GHS, 2017). However, the national HIV prevalence rate has generally increased in recent years (Table 2.4). The HIV population among young adults 15–24 years also increased between 2015 (36,560) and 2019 (37,423), representing a 2.4% increase (GHS and NACP, 2019).

In Ghana, in 2018 for instance, grouping prevalence rates by age and location, adolescents (15–24) recorded the highest prevalence rate, in both rural and urban areas. The youth had 32.9% prevalence, the highest for age groups in the country in 2018. Adolescents accounted for the highest HIV incidence rate in the rural

Table 2.4 Trends in HIV Prevalence (%) by Age Group, 2008–2018

Age Group	*2008*	*2009*	*2010*	*2011*	*2012*	*2013*	*2014*	*2015*	*2016*	*2017*	*2018*
15–19	1.2	1.9	1.1	1.9	0.7	0.8	0.9	0.7	0.6	0.9	0.9
20–24	2.3	2.2	1.5	1.5	1.5	1.5	1.4	1.4	1.3	1.8	1.8
25–29	2.8	3.7	2.5	2.7	2.8	2.4	2.0	2.0	2.1	2.3	2.3
30–34	2.8	3.4	2.8	2.9	3.0	2.5	2.8	2.9	3.3	2.4	3.0
35–39	2.8	3.6	2.8	2.8	3.3	3.2	3.2	3.4	3.5	3.4	3.4
40–44					3.2	2.7	2.1	2.6	2.6	3.4	3.0
45–49						3.3	1.7	1.9	5.7	3.2	1.6
15–24[1]	1.9	2.1	1.5	1.7	1.3	1.2	1.8	1.1	1.1	1.5	1.5
National	2.2	2.9	2.0	2.1	2.1	1.9	1.6	1.8	2.4	2.1	2.4

Sources: Global Health Data Exchange (2020); GHS (2017, 2018); WHO (2015)

Note: 1 Median values

communities. In the urban areas, infection in adolescents constituted the second highest (GHS and NACP, 2019). This increase is attributed to complacency in the country's ability to previously lower its national prevalence rate (Table 2.4). It can also be partly attributed to reduction in public campaigns on HIV over the period (GSS, GHS and ICF, 2015).

Adolescents with HIV in Ghana experience stigma. They also have conflicting views about clinic attendance and antiretroviral therapy. Those who muster courage to attend clinic experience serious discrimination (Adinkrah, 2017; Dako-Gyeke et al., 2020). The need for psychosocial interventions and financial assistance to adolescents in Ghana living with HIV and their families has also been highlighted (Dako-Gyeke et al., 2020).

Sexual transmission of HIV is the predominant mode of transmission in Ghana, and heterosexual sex is the main mechanism of sexual transmission. Mother to child (vertical) transmission is an important mode of HIV transmission and is the most common route of HIV infection in children less than 14 years of age. Sharing injection drug equipment is also recognized as an important mode of HIV transmission among people who inject drugs (GHS and NACP, 2019).

As is common with sub-Saharan Africa (SSA), identifiable population groups in Ghana have higher rates of the infection (Ramjee and Daniels, 2013). Population groups with HIV prevalence higher than the 2014 GDHS adult population average of 2% are: men who have sex with men (MSM) – 18.1% of an estimated total population size of about 55,000 (GAC, 2017), female sex workers (FSWs) – 6.9% of an estimated total population size of about 65,000 (GAC, 2015). Patients with STIs – 6.3% (GAC, 2018); tuberculosis (TB) patients – 14.7%; prisoners – 2.3% (GAC, 2013); and pregnant women – 2.1% (GHS and NACP, 2019).

Key populations (KPs) in Ghana include MSM, FSWs, persons who inject drugs (PWIDs), and prisoners (Table 2.5). KPs are disproportionately affected by HIV in the country. The total number of KPs in the country is not accurately

Table 2.5 HIV/AIDS incidence by sex, age, and key populations

Indicator	*Year*											
	2008	*2009*	*2010*	*2011*	*2012*	*2013*	*2014*	*2015*	*2016*	*2017*	*2018*	*2019*
HIV incidence[1] (national)	0.14	0.14	0.13	0.13	0.12	0.12	0.12	0.11	0.11	0.11	0.11	0.1
HIV[2] incidence by sex:												
Male	0.1	0.1	0.09	0.09	0.08	0.08	0.08	0.08	0.08	0.08	0.08	0.07
Female	0.18	0.18	0.17	0.17	0.16	0.15	0.15	0.15	0.15	0.14	0.14	0.14
HIV incidence by age[3]												
15–19	0.9	0.89	0.86	0.83	0.79	0.76	0.75	0.74	0.73	0.71	0.70	0.68
20–24	1.7	1.68	1.61	1.55	1.47	1.42	1.41	1.39	1.37	1.33	1.31	1.26
25–29	1.9	1.87	1.8	1.73	1.64	1.59	1.57	1.54	1.51	1.48	1.45	1.4
30–34	1.71	1.69	1.63	1.57	1.49	1.45	1.43	1.4	1.38	1.34	1.32	1.27
35–39	1.41	1.4	1.35	1.29	1.23	1.19	1.18	1.16	1.15	1.12	1.1	1.05
40–44	1.12	1.09	1.04	1.01	0.96	0.94	0.94	0.92	0.9	0.88	0.87	0.83
45–49	0.87	0.86	0.83	0.81	0.76	0.73	0.72	0.7	0.69	0.68	0.67	0.65
15–24	0.13	0.13	0.12	0.12	0.11	0.11	0.11	0.10	0.10	0.10	0.10	0.10
HIV incidence by key populations (/1000)	0.97	0.97	0.91	0.88	0.83	0.82	0.8	0.79	0.75	0.72	0.7	0.66
FSWs	0.03	0.03	0.03	0.03	0.02	0.02	0.02	0.02	0.02	0.02	0.02	0.02
Clients of FSWs	0.05	0.05	0.05	0.04	0.04	0.04	0.04	0.04	0.04	0.04	0.03	0.03
Partners or Clients of FSW	0.10	0.10	0.10	0.09	0.09	0.09	0.08	0.08	0.08	0.08	0.07	0.07
MSM	0.03	0.03	0.03	0.03	0.03	0.03	0.03	0.03	0.03	0.03	0.03	0.02

Source: Author, 2020, calculated on the basis of Modes of Transmission (MOT) survey (GAC, 2014)

Notes: 1 New infections (incidence, %), for ages 15–49

2 Incidence, %.

3 Incidence per 1000 for age groups

known, but this is estimated to be less than 1% of Ghana's population of nearly 30 million people. However, FSWs and their clients, and MSM, account for 28% of all new HIV infections (GHS and NACP, 2019). Residents of port and boarder communities also typically have high HIV infection. In 2019, for instance, the main harbor city of Tema recorded 1,072 new cases (Tema Port Health Authority/GhanaWeb, 2020). Target 3 of the SDG3 stipulates ending the HIV epidemic by 2030 (Inter-Agency and Expert Group on SDG Indicators [IAEG-SDGs], 2016, p. 4). The current trend of the infection casts a shadow on the feasibility of Ghana achieving this target by 2030.

Gendered dynamics of HIV/AIDS in Africa and Ghana

Adolescents in Africa have disproportionate proportions of new HIV-positive infections, with 73% in 2016, for example (UNICEF, 2016). Moreover, HIV/AIDS is gendered globally including in SSA, with women bearing a disproportionate burden of it (Kharsany and Karim, 2016; Ramjee and Daniels, 2013). Young women are particularly burdened; girls form 75% of new infections in SSA among persons aged 15–19 (UNAIDS, 2019). A young woman is said to be infected every passing minute (UNAIDS, 2012). Similarly, in Ghana, HIV/ADIS is gendered (GHS and NACP, 2019; Owusu, 2020), particularly for adolescents. For instance, of the 2018 incidence of 334,713, 65% were females (GHS and NACP, 2019).

Based on the nationally representative 2003 GDHS data, Sia et al.'s (2016) analysis unearthed critical HIV/AIDS gender differentials. A gender disparity of 1.08 was found for women. 92% of the higher female HIV/AIDS prevalence, compared to males, was significantly explained by age at sexual debut. Also, the study found that gendered differences in sexual behavior explained 47.1% of the excess HIV prevalence for females. For example, if their female samples reported being virgins, 15.5% of the gender differential was explained, while 24% was explained if their male respondents mentioned being virgins.

Having the same level of virginity would decrease the gender differentials in HIV seropositivity in Ghana by 57.4% (Sia et al., 2016). The differential distribution of premarital sex for males and females (54.3% and 43.1%, respectively) reduced the gender gap in HIV/AIDS prevalence (Sia et al., 2016). Furthermore, 38.6% of excess HIV/AIDS infection in women was explained by the gendered differentials in the distribution of marital status: widowhood, divorce, or separation was particularly important. Thus, Ghana's inability to achieve the MDG targets regarding HIV/AIDS disadvantages women the more.

In recent years, both the incidence and the prevalence rates of HIV/AIDS in Ghana give alarming information: adolescents make up a large chunk of these numbers. Undoubtedly this has negative implications for national development as the robust and energetic human resource needed to contribute to productivity and undertake roles in nation building is getting infected by HIV. More worrying, survival of HIV/AIDS remains a major concern among young people (Cluver et al., 2018). Adolescents' rates of adherence and retention to care remain low (Lowenthal et al., 2014; UNICEF, 2016), while viral suppression failure (Ferrand et al., 2016) and TB are common (WHO, 2016).

Reasons for high HIV/AIDS rates in female youth in Ghana

In Ghana, as in SSA generally, a number of factors account for the higher prevalence of adolescent female HIV rates than males. It could largely be attributed to lack of access to education on sexual health, poverty, and gender-based and intimate partner violence. Others include child marriage, transactional sex, including "sugar daddy" culture (where much younger females, particularly adolescent girls, befriend and trade sex with much older and wealthier men for valuable goods), and young women's lack of access to healthcare. Additional reasons include women's lower educational and socio-economic status, which make them vulnerable.

Culturally motivated marital norms are also important in this respect and increase the spread of HIV. In SSA, cultural norms encourage younger females to be married to older males (Koski, Clark and Nandi, 2017). Polygamy and extra marital affairs are permitted for males but shunned for females (Anarfi and Owusu, 2011; Sedziafa, Tenkorang and Owusu, 2016). Also, sex workers are usually adolescent females/young women, hence facilitating the spread of the infection among young females in Ghana. Poverty coupled with low literacy, economic dependency on men, and lack of higher education are additional reasons. Lack of appreciable employment, archaic cultural practices, which put women at sexual risk, and women's lack of control of their sexuality, particularly adolescents/young women, have further made them more vulnerable and at a higher risk of contracting the HIV infection (Sedziafa, Tenkorang and Owusu, 2016).

Furthermore, in Ghana, these high prevalence and infection rates among young people, especially females, can be attributed to the low levels of knowledge about HIV/AIDS prevention among adolescents aged 15–24 years. As at 2018, only 22.15% of the youths were knowledgeable in HIV prevention (Hope for Future Generations, 2019). Education and awareness on HIV among female adolescents is lower than that of male adolescents. Overall, females are less likely than males to have comprehensive knowledge on HIV (GSS, GHS and ICF, 2015). In 2014, awareness of HIV prevention for females was 19.9% while that of men was 27.2% for nationally representative samples (GSS, GHS and ICF, 2015). The higher HIV prevalence rates in adolescent females than males are also accounted for by the fact that generally, males in Ghana have higher education on HIV prevention methods than females. For example, in Ghana, more males than females had a higher knowledge of sources to procure a condom, which often facilitates procuring condoms. Eighty-eight percent of men versus 72% of women knew at least one condom source (GSS, GHS and ICF, 2015).

A significant decrease in the numbers of young females infected with HIV in Ghana and SSA will generally require social reforms that will empower and improve the livelihood of young women. These will include encouraging females to take up more and higher educational opportunities. Females should be encouraged to study more technical subjects to help change the situation on the formal job market in Ghana and Africa generally, where more technical jobs which usually pay higher remuneration are mostly the preserve of males. Females should also be encouraged and supported to vie for and ascend more managerial positions. UN Women's (2018) response to achieving SDG5 includes ensuring gender equality to economic resources, and availing social protection and economic stimulus mitigation efforts which positively impact women and girls. The need for data and coordination efforts that will encompass gender perspectives has also been proposed. As this chapter has unequivocally demonstrated, the call for availing universal access to sexual and reproductive health (UN Women, 2018) is particularly important for young women and girls in Ghana.

Laws addressing the issues of gender inequality, violence, and exploitation against women (UN Women, 2018) should also be enacted and monitored with the necessary punitive measures taken. School-based interventions should be enforced to ensure that young adolescent females receive age-appropriate education on their sexuality, reproductive health, use of contraception, and safer negotiation methods in romantic relationships. Efforts at women's empowerment should also be rooted at the community levels. Among others, community opinion leaders should be carefully groomed to root out and prevent the entrenched social norms that work against females. Inheritance norms which discriminate against women, for instance, should be changed. The selection, grooming, and support of more female opinion leaders to play political, religious, and social roles should be encouraged at local, regional, and national levels. The efforts of some Ghanaian females in high-level leadership positions who are grooming young females to model them are applauded. These eminent persons include Her Ladyships Akua Kuenyehia and Georgina Theodora Woode. Kuenyehia was elected a judge at the International Criminal Court between 2003 and 2015, and served as its First Vice-President. Woode was Ghana's 12th Chief justice, from June 2007 to June 2017, and the country's first woman Chief Justice.

Challenges

Implementation of the SDGs, particularly in Ghana, is confronted with some challenges. These include data availability, particularly age and sex-disaggregated data for male and female adolescents/young persons. Furthermore, in Ghana, available data may not be presented at internationally required levels. For instance, the incidence of malaria, TB, health insurance coverage, etc. is yet to be measured per 1,000 population, as the SDG international indicators stipulate for monitoring and evaluation of progress. The required frequency of such data, for instance, on annual basis, may also be lagging.

Conclusion

This chapter has focused on the current state of select indicators of SDG3 (maternal, under-five, and neonatal mortality). The SDGs create a unique opportunity for improving the health status and dignity of Ghanaians, particularly, females and adolescents/young people. They particularly illuminate gender equality as a cross-cutting strategy to achieving the SDGs internationally (UN Women, 2018). As highlighted by this chapter, promoting gender equality for women and girls in Ghana, by improving their health and access to health services, is an unequivocal element through which the SDGs will yield equity, inclusiveness, poverty alleviation, tranquility, protection, and sustainable development (UN Women, 2018). The chapter also highlighted some contextual and cultural tenets. It has drawn attention to policy, legal, and implementation vacuum/lapses which mitigate against the human rights of females and young girls particularly. Data vacuums have also been emphasized.

To achieve the SDGs, policy analysts point to the need to focus on regional and local policy contexts by specifically considering the budget prioritization and fiscal capacity of signatory countries as these remain key challenges to the implementation of the global goals (Moldalieva et al., 2016; Konadu and Osei, 2020). Setting global goals alone might not be enough for real change to occur. In addition to the budgetary and policy arenas, this chapter has demonstrated the critical need to work on changing entrenched cultural practices that militate against the Ghanaian society generally, and in particular the youth, especially females. The chapter has shown that given the current state of affairs regarding the select SDG3 targets and indicators vis-a-vis prevalent logistical, policy, and legal challenges, achieving these targets is outside the remit of Ghana.

It is recommended that stakeholders pay attention to poverty alleviation, enforce existing policies, and retool the law enforcement and criminal justice systems. These would include a critical analysis of fiscal capacity to back the realization of the SDGs (Moldalieva et al., 2016; Konadu and Osei, 2020). Local opinion leaders should be educated to expunge archaic and dangerous societal norms and practices. Females and girls should particularly be protected in a bottom-up approach. Communities should groom and showcase female community members who have exceled socially to boost the image and dignity of females. Quality data, which are disaggregated by age cohorts, sex, socio-demographic quintiles, geographic, and other pertinent socio-economic indices, and meet international targets, are an urgent necessity for informed decision-making, monitoring, and evaluation. Health and legal advocates should liaise more with traditional leaders to ensure mitigation of community and national factors that militate against achieving SDG3, particularly for Ghana's teeming youth. Finally, a cross-cutting emphasis on promoting gender equality for young women and girls in Ghana is recommended.

References

Adinkrah, M. (2017) 'When a Wife Says "No": Wife Sexual Refusal as a Factor in Husband: Wife Homicides in Ghana', *Journal of Interpersonal Violence*, 36(3–4), pp. 1607–1633. doi: 10.1177/0886260517742913.

African Union (2006) *African Youth Charter*. Banjul, Gambia: Africa Union, Gambia.

Ahiadeke, C. (2001) 'Incidence of Induced Abortion in Southern Ghana', *International Family Planning Perspectives*, pp. 96–108.

Amankrah, J. Y. (2003) *Ghana: Decent Work Statistical Indicators: Fact Finding Study*. Geneva: International Labour Organization.

Amissah, C. M. and Nyarko, K. (2017) 'Psychological Effects of Youth Unemployment in Ghana', *Journal of Social Sciences*, 13(1), pp. 64–77. doi: 10.3844/jssp.2017.64.77.

Anarfi, J. K. and Owusu, A. Y. (2011) 'The Making of a Sexual Being in Ghana: The State, Religion and the Influence of Society as Agents of Sexual Socialization', *Sexuality & Culture*, 15(1), pp. 1–18.

Blaboe, N. A. (2019) *Assessing SDG 3: Achieving Universal Health Coverage in Ghana*. Accra, Ghana: University of Ghana.

Boah, M., Bordotsiah, S. and Kuurdong, S. (2019) 'Predictors of Unsafe Induced Abortion among Women in Ghana', *Journal of Pregnancy*. https://doi.org/10.1155/2019/9253650

Chongcharoentanawat, P. *et al.* (2016) 'The Affordability of the Sustainable Development Goals: A Myth or Reality?', *United Nations University-Maastricht Economic and Social Research Institute on Innovation and Technology Working Paper Series.*

Cluver, L. *et al.* (2018) 'Sustainable Survival for Adolescents Living with HIV: Do SDG-Aligned Provisions Reduce Potential Mortality Risk?', *Journal of the International AIDS Society*, 21, p. e25056.

Dako-Gyeke, M. *et al.* (2020) 'Understanding Adolescents Living with HIV in Accra, Ghana', *Children and Youth Services Review*, 108, p. 104590.

Domfe, G. and Oduro, A. D. (2018) *Prevalence and Trends in Child Marriage in Ghana.* Centre for Social Policy Studies, College of Humanities University of Ghana. Accra, Ghana: University of Ghana.

FAO (2005) *AQUASTAT FAO's Global Information System on Water and Agriculture.* Rome: FAO.

Ferrand, R. A. *et al.* (2016) 'Viral Suppression in Adolescents on Antiretroviral Treatment: Review of the Literature and Critical Appraisal of Methodological Challenges', *Tropical Medicine & International Health*, 21(3), pp. 325–333.

GAC (2013) *Integrated Biological and Behavioral Surveillance Survey (IBBSS) 2013.* Accra, Ghana: Ghana AIDS Commission.

GAC (2014) *Country AIDS Response Progress Report – Ghana.* Accra: Ghana AIDS Commission.

GAC (2015) *Integrated Biological and Behavioral Surveillance Survey (IBBSS).* Accra, Ghana: Ghana AIDS Commission.

GAC (2017) *Ghana Men's Study (2017): Mapping and Population Size Estimation (MPSE) and Integrated Bio-Behavioral Surveillance Survey (IBBSS) amongst Men Who Have Sex with Men (MSM) in Ghana (Round II).* Accra, Ghana: Ghana AIDS Commission. Available at: https://www.ghanaids.gov.gh/mcadmin/Uploads/Ghana%20Men's%20Study%20Report(2).pdf (Accessed 11 March, 2020).

GAC (2018) *Global AIDS Monitoring Report – Ghana.* Accra, Ghana: Ghana AIDS Commission. Available at: https://www.unaids.org/sites/default/files/country/documents/GHA_2018_countryreport.pdf (Accessed 29 February, 2020).

GAC (2019) *National HIV and AIDS Policy: Universal Access to HIV Prevention, Treatment and Care Services Towards Ending AIDS as a Public Health Threat.* Accra, Ghana: Ghana AIDS Commission. Available at: https://ghanaids.gov.gh/mcadmin/Uploads/nationalHIVandAIDSPolicy.pdf

Ghana Health Service (GHS) (2017) '2016 Annual Report of the ABPN', *The American Journal of Psychiatry*. doi: 10.1176/appi.ajp.2017.174804.

Ghana Health Service (GHS) (2018) *The Health Sector in Ghana: 2017 Facts and Figures.* Accra: Ministry of Health. Ghana.

Ghana Statistical Service (2014) *Ghana Living Standards Survey Round 6 (GLSS 6): Poverty profile in Ghana (2005–2013).* Accra: Ghana Statistical Service.

Ghana Statistical Service (GSS), Ghana Health Service (GHS) and ICF (2015) *2014 Ghana Demographic and Health Survey (DHS) Key Findings, GSS, GHS, and ICF International*, pp. 4–6. Accra: Ghana Statistical Service.

Ghana Statistical Service (GSS), Ghana Health Service (GHS) and ICF (2018) *Ghana Maternal Health Survey 2017.* Accra: Ghana Statistical Service.

GHS and NACP (2019) *2018 HIV Sentinel Report.* Accra: Ministry of Health. Ghana. Global Health Data Exchange (2020) *Ghana Country Data Profiles.* Seatle: Institute for Health Metrics and Evaluation, Seatle: University of Washington. Available at: http://ghdx.healthdata.org/geography/ghana. (Accessed 16 April, 2020).

Grindlay, K. *et al.* (2018) 'Contraceptive Use and Unintended Pregnancy among Young Women and Men in Accra, Ghana', *PLoS One*, 13(8), p. e0201663.

GSS (2012) *2010 Population and Housing Census, Summary of Report of Final Results.* Accra: GSS, pp. 1-117.

Hashiguchi, L. (2016) *Positioning Adolescent Health as a Cornerstone of the SDGs*. Available at: file:///C:/Users/Shalom/Desktop/HIV LIT REVIEW/Positioning adolescent health as a cornerstone of the SDGs Institute for Health Metrics and Evaluation.html. (Accessed 7 March 2020).

Hope for Future Generations (2019) *Involve Youth in HIV Interventions*. Available at: https://hffg.org/involve-youth-in-hiv-interventions-hffg/. (Accessed 16 January 2020).

Inter-Agency and Expert Group on SDG Indicators [IAEG-SDGs] (2016) *Inter-Agency and Expert Group on SDG Indicators [IAEG-SDGs].*

ISSER (2017) *Ghana Social Development Outlook 2016*. Accra. The Institute of Statistical, Social and Economic Research. University of Ghana.

Karver, J., Kenny, C. and Sumner, A. (2012) 'MDGs 2.0: What Goals, Targets, and Timeframe?', *IDS Working Papers*, 2012(398), pp. 1–57.

Kharsany, A. B. and Karim, Q. A. (2016) 'HIV Infection and AIDS in Sub-Saharan Africa: Current Status, Challenges and Opportunities', *The Open AIDS Journal*, 10, 34–48. Epub 2016/06/28. https://doi. org/10.2174/1874613601610010034 PMID: 27347270.

Konadu, O. A. and Osei, V. K. (2020) 'Can Ghana Afford the Sustainable Development Goal on Education?', *Africa Education Review*, 17(2).

Koski, A., Clark, S. and Nandi, A. (2017) 'Has Child Marriage Declined in Sub-Saharan Africa? An Analysis of Trends in 31 Countries', *Population and Development Review*, pp. 7–29.

Lowenthal, E. D. *et al.* (2014) 'Perinatally Acquired HIV Infection in Adolescents from Sub-Saharan Africa: A Review of Emerging Challenges', *The Lancet Infectious Diseases*, 14(7), pp. 627–639.

Lund, R. and Agyei-Mensah, S. (2008) 'Queens as Mothers: The Role of the Traditional Safety Net of Care and Support for HIV/AIDS Orphans and Vulnerable Children in Ghana', *GeoJournal*, 71(2–3), pp. 93–106.

Lutz, W. *et al.* (2019) 'Education Rather Than Age Structure Brings Demographic Dividend', *Proceedings of the National Academy of Sciences*, 116(26), pp. 12798–12803.

MOH (2017) *Medium Term Expenditure Framework for 2018–2021: Program Based Budget Estimate for 2018*. Available at: www.mofep.gov.gh/sites/default/files/pbb-estimates/2018/2018-PBB-MoH.pdf. (Accessed 12 February 2020).

MOH and UNDP (2011) *Leveraging Fiscal Space for Human Development in Ghana. The 2015 MDG Targets and Beyond*. Accra: Government of Ghana.

Moldalieva, J., Muttaqien, A., Muzyamba, C., Osei, D., Stoykova, E., and Le, N. (2016) *Millennium Development Goals (MDGs): Did they change social reality? (No. 2016-035)*. Maastricht, The Netherlands: United Nations University-Maastricht Economic and Social Research Institute on Innovation and Technology (MERIT).

NACP (2009) *Ghana HIV Sentinel Survey. Dataset Records for National AIDS Control Program (Ghana)*. Accra, Ghana: National AIDS Control Program. Avialable at: http://ghdx.healthdata.org/organizations/national-aids-control-program-nacp-ghana

NDPC and UN Ghana (2015) *Ghana Millenium Development Goals, United Nations Development Program*. Available at: www.gh.undp.org/content/dam/ghana/docs/Doc/Inclgro/UNDP_GH_2015GhanaMDGsReport.pdf. (Accessed 18 March, 2020).

Nii-Boye Quarshie, E. *et al.* (2015) 'Adolescent Suicide in Ghana: A Content Analysis of Media Reports', *International Journal of Qualitative Studies on Health and Well-Being*, 10(1), p. 27682.

Owusu, G. A. (2007). Graying of the developing world: emerging policy issues. *Legon Journal of International Affairs*, *4*(1), 1–25.

Owusu, A. Y. (2020). A gendered analysis of living with HIV/AIDS in the eastern region of Ghana. *BMC Public Health*, 20, 75, DOI: 10.1186/s12889-020-08702-9.

Payne, C. M. *et al.* (2013) 'Why Women Are Dying from Unsafe Abortion: Narratives of Ghanaian Abortion Providers', *African Journal of Reproductive Health*, 17(2), pp. 118–128.

Ramjee, G. and Daniels, B. (2013) 'Women and HIV in Sub-Saharan Africa', *AIDS Research and Therapy*, 10(1), pp. 1–9.

Republic of Ghana (2004) *Republic of Ghana and Ministry of Women and Children's Affairs*. Accra, Ghana: Government of Ghana.

Republic of Rwanda (2019) *2019 Rwanda Voluntary National Review (VNR) Report, Sustainable Development Goals* (June), pp. 1–109.

Rominski, S. D. and Lori, J. R. (2014) 'Abortion Care in Ghana: A Critical Review of the Literature', *African Journal of Reproductive Health*, 18(3), pp. 17–35.

Sedziafa, A. P., Tenkorang, E. Y. and Owusu, A. Y. (2016) '". . . He Always Slaps Me on My Ears": The Health Consequences of Intimate Partner Violence among a Group of Patrilineal Women in Ghana', *Culture, Health & Sexuality*, 18(12), pp. 1379–1392.

Sheehan, P. *et al.* (2017) 'Building the Foundations for Sustainable Development: A Case for Global Investment in the Capabilities of Adolescents', *The Lancet*, 390(10104), pp. 1792–1806.

Sia, D. *et al.* (2016) 'What Explains Gender Inequalities in HIV/AIDS Prevalence in Sub-Saharan Africa? Evidence from the Demographic and Health Surveys', *BMC Public Health*, 16(1), pp. 1–18.

Tema Port Health Authority/GhanaWeb (2020) *1,072 Test HIV Positive in Tema*. Available at: www.ghanaweb.com/GhanaHomePage/NewsArchive/1-072-test-HIV-positive-in-Tema-862264. (Accessed 8 March 2020).

UNAIDS (2012) *Every Minute, a Young Woman Is Infected with HIV*. Geneva, Switzerland: Joint United Nations Program on HIV/AIDS (UNAIDS). Available at: www.unaids.org/sites/default/files/media_asset/UNAIDS_Gap_report_en.pdf.

UNAIDS. (2019) *Global HIV & AIDS Statistics: 2019 Fact Sheet*. Available at: www.unaids.org/en/resources/fact-sheet. (Accessed 4 February 2020).

UNDP, Ghana and NDPC/GOG (2012) *2010 Millennium Development Goals Report*.

UNICEF (2016) *For Every Child, End AIDS: Seventh Stocktaking Report, 2016*. New York: UNICEF.

United Nation (no date) *Sustainable Development Goals: Goal 5: Achieve Gender Equality and Empower All Women and Girls*. Available at: www.un.org/sustainabledevelopment/gender-equality/. (Accessed 15 October 2020).

United Nations (1981) *Definition of Youth: Secretary General's Report: General Assembly Report (A/36/215) and Resolution 36/28, 1981*. New York: United Nations.

UN Women (2018) 'Why gender equality matters across all SDGs. Turning promises into action: Gender equality in the 2030.' p. 33.

WHO (2014) 'World Health Organization: Health for the World's Adolescents a Second Chance in the Second Decade', *World Health Organization.*

WHO (2015) *HIV Country Profile – Ghana*. Geneva: World Health Oorganization.

WHO (2016) *Global Tuberculosis Report*. Geneva: World Health Organization.

WHO (2017a) *Global Strategy for Women's, Children's and Adolescents' Health (2016–2030): Adolescents' Health*. Geneva: WHO. (A70/37). Available at: apps.who.int/gb/ebwha/pdf_files/WHA70/A70_37-en.pdf. (Accessed 16 January 2020).

WHO (2017b) *Global Accelerated Action for the Health of Adolescents (AA-HA!): Guidance to Support Country Implementation*. Geneva. Available at: www.who.int/maternal_child_adolescent/topics/adolescence/framework-accelerated-action/en. (Accessed 16 January 2020).

WHO (2018) *Engaging Young People for Health and Sustainable Development: Strategic Opportunities for the World Health Organization and Partners*.

World Population Review (2020) *Ghana Population 2020*. Available at: http://worldpopulationreview.com/countries/ghana-population/. (Accessed 9 March 2020).

3 Sustainable Development Goals and the internal logics of "gender equality" in the Liberian context

Erica S. Lawson, Florence W. Anfaara, and Ola Osman

Introduction

This chapter addresses the internal logics of "gender equality" in the Liberian context as it pertains to Sustainable Development Goal 5 (SDG5). Its purpose is to examine the interpretive politics of gender equality as articulated by Liberian women on the basis of their experiences in navigating political, local, and international systems that simultaneously hinder and present opportunities for gender equality. Stated in the United Nations' most recent iteration of Development Goals, SDG 5 seeks to "Achieve gender equality and empower all women and Girls." Furthermore, The UN explains that: "Gender equality is not only a fundamental human right, but a necessary foundation for a peaceful, prosperous and sustainable world" (Sustainable Development Goals, 2021).

Likewise, in our collaboration with Liberian women, they note that persistent gender inequality is detrimental to the country's peacebuilding and transitional justice initiatives; they argue that they face challenges in protecting fragile peace gains since the end of the 14-year civil war (1989–2003), and express continued commitment to working with international institutions and local organizations to promote gender equality within a peacebuilding and justice framework. Put differently, Liberian women must navigate a challenging political and socioeconomic landscape anchored in patriarchal practices in a post-conflict country ruled by an elite class. At the same time, they live in a country located in a global economy where local resources are extracted for profit at the expense of ordinary Liberians with greater burdens on women's social reproductive labor.

At the heart of the vision for gender equality is the socially desirable "empowered" woman who, it is argued, will lift herself, her family, and community out of poverty, if presented with the right educational or financial opportunities (e.g., access to credit). While it is beyond the scope of this chapter to fully address, feminist scholars have provided insightful critiques about the "empowered woman" at the center of "development," to the extent that such a woman is held responsible for advancing progress in her society (Celis et al., 2008; Duflo, 2012; Kabeer, 2015), rather than addressing structural inequalities that continue to hamper the poorest "postcolonial" countries who are exploited to support Western economies. Although "empowerment" is contested within feminist discourses on

DOI: 10.4324/9780367743918-4

"development" (Khurshid, 2016; Khoja-Moolji, 2017), at its core is the neoliberal promise that dislodging African women from "detrimental" cultural and traditional practices offers them autonomous and modernist futures. This is a promise fraught with complications within sustainable development goals with their focus on gender equality that may not be deeply informed by the complex and multidimensional realities of the day-to-day lives of African women, or the intracommunal understanding of "human rights" differently envisioned from those that frame Western philosophical and political traditions. In this context, we ask, how do Liberian women conceptualize and articulate gender equality in relation to peace and justice as this is informed by intersectional and daily realities?

In exploring this question, we draw on qualitative interviews conducted with Liberian women between 2017 and 2018. These are women who provide conflict resolution services in locally based Peace Hut councils across the country, who aspire to, and those who occupy political office, as well as women who run local NGOs focused on justice and equality. These interviews were part of a larger research project with Liberian collaborators to document and analyze how women across social sectors, but particularly in local communities, undertake conflict resolution and mediation activities to de-escalate violence and resolve disputes. Women who do this work in Peace Huts are of the view that dispute resolutions led by women in communities can promote gender equality and quell the potential for civil unrest or war.

Toward the central objective, which is to highlight how women understand and articulate gender equality, this chapter is organized as follows: First, we discuss the Millennium Development Goals (MDGs) in Liberia, with a focus on MDG 3, and the context for understanding the new round of SDGs, with particular emphasis on SDG 5 as these pertain to gender equality on the continent and in Liberia. Here, we focus on the progress and setbacks with respect to MDG 3 under the administration of former President Ellen Johnson Sirleaf. Second, we address the African Union's (AU's) support for gender equality in its long-term strategic plan for advancing African progress in the twenty-first century, as well as Liberian women's advocacy for gender justice in a post-war context. Third, we present a fuller explanation of our study methodology and data analysis, followed by a representation and discussion of select narratives based on interviews in which we highlight five central themes that are indicative of women's understanding of gender equality. Finally, we offer concluding remarks about the complexities and contradictions that underpin the conceptualization and advancement of gender equality in Liberia.

Assessing MDG 3 in the African/Liberian context

Gender equality is defined as the "equal rights, responsibilities and opportunities for women and men and girls and boys" (UN Women, 2021; Canadian Women's Foundation, 2020), and is widely seen as a driving force for growth and development. As such, there is a widespread belief that advancing Millennium

Development Goal (MDG 3), and now SDG 5, is key to meeting other sustainable development goals by 2030 (UNESCO, 2017; United Nations MDG Report, 2015). The MDGs were launched in September 2000 following the UN Millennium Summit of world leaders who agreed on eight developmental goals targeted at reducing global poverty, as well as promoting human rights and justice (Hulme, 2007; Kabeer, 2015). According to Naila Kabeer (2015), MDG 3 was meant to be the pillar for all other goals such that centering gender equality would effectively "combat poverty, hunger and disease and . . . stimulate development that is truly sustainable" (Kabeer, 2015, p. 383).

A post-MDG assessment undertaken in 2015 revealed that on a global scale, the world had achieved most of the MDG targets including gender equality. According to the United Nations MDG Report (2015), the poverty rate in many developing countries was reduced by 33%, suggesting a drastic change from people living on less than $1.00 a day to people now living on at least $1.25 a day; gender discrimination in primary and secondary schools was addressed with the increased enrollment of girls; and, additionally, about 41% of women now work in paid employment other than the agricultural sector. Furthermore, according to the report, the number of women representatives in parliaments across the continent had doubled by 2015.

However, the 2015 UN Development Programme report also noted that Liberia was only able to meet three of the eight MDGs in the areas of gender equality, HIV/AIDS, and global partnership. Under the administration of President Ellen Johnson Sirleaf (2006–2017), the three markers of gender equality – education, employment, and political representation – were tackled but not adequately achieved. First, the ratio of female to male school enrollment increased from 72% in 2000 to 90% in 2009 at the primary level, and from 71% to 75% at the secondary level. Improvements were also achieved in the areas of macroeconomic stability, governance and the rule of law, commitment to fiscal responsibility, and increased engagement with strategic development partners (Hanna and Alfaro, 2012, p. 78).

Second, there was overall improvement in the country's economy. Notably, prior to 2009, Liberia ranked at the bottom of world rankings for human capacity building. However, the country improved to rank at 169 out of 182 countries listed in the UN Human Development Index (Hanna and Alfaro, 2012). Among others, this improvement is indicative of measures taken by the Johnson Sirleaf administration to support market women. Specifically, the government expanded and renovated local markets to cater to the vast majority of Liberian ("market") women who work in the informal sector. Also, a small-scale loan scheme was established to support market women who wanted to participate in transnational trade (Fleshman, 2010). Furthermore, women's voices were loudest during Johnson Sirleaf's administration as a result of the modest political visibility she provided for them.

Third, under her administration, the Women's Legislative Caucus was established to focus attention on gender-sensitive legislative proceedings. For example, women's representation increased from 6% in 1995 to about 13% in 2010 in the national legislature and 17% in the senate (Hanna and Alfaro, 2012, p. 77). The administration also increased female political appointments in finance, commerce and industry, and foreign affairs, including female ambassadorial appointments to South Africa, Scandinavian countries, and Germany (Tulay-Solanke, 2018).

Despite these efforts, the country still lags behind in achieving political and economic equality for women. Not only do Liberian women lack political support by the state, they lack community support as well. In the recent 2020 senatorial elections, only two out of the 18 women parliamentary aspirants were elected into parliament (National Election Commission, 2020), a modest accomplishment resulting from a hard fight. For example, in Gbarpalu County, where one of the female parliamentarians was elected, the paramount chief of the area is reported to have snatched ballot boxes at four polling stations and invoked the *country devil*[1] to prevent women from voting, especially for the female candidate (Geterminah, 2020). Here, we see how patriarchal practices continue to deny women full political participation.

Although no longer in power, Johnson Sirleaf is criticized by Liberian feminists and other women leaders for her lackadaisical posture in advancing gender equality and laying the groundwork for more women to occupy political office. Indeed, Pailey and Williams (2017) argue that President Johnson Sirleaf disappointed Liberian women with respect to advancing gender equality in four ways.

First, she is criticized for her seemingly unenthusiastic attitude toward the gender equity bill that was presented in the legislature and blamed for its failure to pass. Second, she is accused of supporting more male than female candidates in her Unity Party. Third, she has been criticized for the ineffectiveness of the justice system to prosecute lawmakers who have been accused of rape (Pailey and Williams, 2017); and fourth, she is blamed for neglecting the needs of women who live in rural areas. Notwithstanding these criticisms, Liberian women enjoyed relative political visibility compared to the current administration under President Weah (Tulay-Solanke, 2018) despite his self-declaration as "Feminist-in-Chief" (Executive Mansion, 2018).

Building on the strengths and limitations of the MDGs, on September 2015, the Sustainable Development Goals (SDGs) were adopted by UN member states to end extreme poverty by 2030. The SDGs have 17 goals and 169 targets aimed at achieving the *unfinished business* of the MDGs, and to ensure a just and sustainable world (Lomazzi, 2014). Although the SDGs have a more ambitious outlook than the MDGs, their goals and targets are somewhat more representative of the global development agendas of both rich and poor countries. Shettima (2016) points out that the deliberation process was more consultative and transparent than with the MDGs, the latter which were only targeted at developing countries. For instance, many African leaders, including former Liberian president Johnson Sirleaf, were part of the formulation process of the SDG goals and targets, which is different from how the MDGs were formulated (Kabeer, 2015). In addition, SDG 5 has about seven targets with clear-cut explanations of how to achieve each one. SDG 5 seeks to end all forms of discrimination against women and girls, including an end to violence against them, forced labor and marriage, ensuring access to reproductive health, economic and political rights, as well as prioritizing and valuing women's social reproductive labor at the local and national levels (Shettima, 2016).

Although the SDGs are recent, with some African countries yet to implement strategies to meet them, advancing the goals will be particularly challenging for Liberia, a country that struggled to contain Ebola, and now, Covid-19. As well, Nwafor (2019) notes that Liberia depends on donor support to grow its economy; therefore, with the current Covid-19 induced financial crisis facing Western countries, the future for meeting the SDGs in Liberia looks daunting.

Charting a way forward: a continental approach toward gender equality

Sub-Saharan Africa did not meet most of the MDGs and there is a fear that the SDGs will also be missed (Shettima, 2016). However, the continent is making progress in its efforts to create the *Africa We Want* (African Union, 2021). In May 2013, "Agenda 2063" was adopted by African heads of state to project a 50-year "master plan for transforming Africa into a global powerhouse of the future" (African Union, 2021). "Agenda 2063" aligns with meeting and exceeding the SDGs; for example, aspiration 6 of "Agenda 2063" calls for an "inclusive society where all its members are engaged in decision makings in all aspects where no woman, man or child is left behind" (African Union, 2021). To achieve these goals, the AU developed the Gender Equality and Women's Empowerment (GEWE) framework to make women visible in Africa's development. In addition, the Maputo Protocol (African Charter on the Rights of Women in Africa) seeks to promote the rights of women by encouraging African states to combat all forms of discrimination (African Union, 2021), while the Women, Gender and Development Directorate of the AU (located in Addis Ababa, Ethiopia) is in charge of coordinating the AU's efforts in achieving gender equality and empowerment for African women.

In conjunction with the AU's commitment to advancing gender equality, Liberian women are at the forefront of advocating for human and women's rights outlined in the Liberian Constitution, as well as those enshrined in regional and international charters, with notable success amid a series of setbacks and precarity in the sociopolitical landscape. Liberian Women's postwar struggles to achieve gender equality and to end all forms of discrimination are evident in the work and ideas of the Liberian national gender policy, the women's manifesto, and in women-led NGOs/civil societies. The creation of Liberia's Ministry of Gender and Development (MoGD) is one of the many notable achievements since the end of the war. The MoGD advances issues pertaining to women in the areas of education, health, politics, and the economy. Within this Ministry, and with international support, the National Gender Policy (NGP) was launched in 2009. The NGP was designed to uphold the "political will of the Liberian government to eliminate all forms of gender-based discrimination to achieve gender equality" (The Liberian National Gender Policy, 2009, p. 5). Taking an ambitious approach, the NGP intended to eliminate all gender-related problems in Liberia by 2020 by focusing on peace and national security, livelihoods, human rights, and governance (The

Liberian National Gender Policy, 2009, p. 6). However, 2020 has come and gone and gender equality remains elusive.

In addition, in 2005, just before the first post-war election, Liberian women came together to develop a policy framework in the form of a Manifesto with recommendations to improve the plight of women (Johnson, 2016). The Manifesto highlighted important areas of interest for women such as education, women's rights, peace and conflict negotiations, sexual and reproductive health, and economic and political empowerment (Jennings, 2012). The Manifesto was revised in 2016 and again in 2020 to include themes that aim to empower women in the economy, build healthy families and safe communities, and advance women's political rights and leadership (Johnson, 2016). By updating the Manifesto, Liberian women insist on projecting their expectations for empowerment, advancement, and equality (Jennings, 2012), pressing for government accountability, and keeping their needs on the public agenda.

Women's movements in the country, such as the Women's NGO Secretariat of Liberia (WONGOSOL), as well as community-based Peace Huts, continue to play significant roles in the struggles for gender equality. For instance, WONGOSOL organizes education and sensitization workshops on ending violence against women and girls, advocating for an end to rape culture, and calling for the equal participation of both women and men in decision-making at all levels and sectors of society (WONGOSOL, 2020). In 2014, WONGOSOL collaborated with the UN Mission in Liberia (UNMIL) to organize a five-day training workshop focused on "leveling the playing field for women's participation in Liberia's governance." Similarly, the women in Peace Huts collaborate with state institutions (e.g., the police and statutory legal systems) to mediate domestic and gender-based violence at the community level through Peace Huts tribunals. They are guided by the United Nations Security Council Resolution (UNSCR) 1325 in negotiating gender-based conflicts at the grassroots level.

Having provided an overview of challenges and progress with respect to gender equality and development goals, the following sections of the chapter present a discussion of how Liberian women conceptualize "gender in/equality" within the internal logics of contextual realities, with a specific examination of five interconnected themes that encapsulate their perspectives. The themes were gleaned from a series of interviews with select groups of women.

Methodological overview

Over the course of one year (2017–2018), and in collaboration with Liberian women who work primarily in Peace Huts, the three authors conducted in-depth interviews (IDIs, n=17) and focus groups discussions (FGDs, n=5, with an average of eight people per group). Although there were overlapping questions within each method, the FGDs investigated questions focused on maternal activism and women's contribution to Liberia's peace economy, while the IDIs focused on women's advocacy to end gender inequality (Lawson and Flomo, 2020). Given the specific focus of the IDIs on issues of gender equality in Liberia, this chapter draws on the analysis of IDI data to examine how Liberian women frame the contradictions and contestations in their advocacy for gender equality. Working with our research partners who are deeply embedded in women's organizations across a number of sectors, we recruited and interviewed women leaders in politics, government, media, in local and international NGOs, and in women-led community groups.

The in-person interviews allowed us to gather more personalized information from our participants in ways that would have been challenging in group settings. Furthermore, since some participants occupied high-profile positions, they would likely have been reluctant to share their personal views and opinions within focus groups, particularly if such views contrasted with those of their respective organizations. On average, the IDIs lasted between 30 minutes and an hour and participants consented to being audio-recorded. Questions focused on how women understood and interpreted gender equality as it relates to their everyday lives; and their assessment of how gender equality can be achieved in Liberia.

Data analysis was undertaken using a mixed inductive–deductive thematic analysis technique. This approach allows data analysis to be driven by both emerging themes and the research's theoretical underpinnings (Antabe et al., 2021). While the researchers identified general themes that emerged from the data, simultaneously, we attended to how the narratives framed gender equality and how the path(s) toward its achievement compared and contrasted with standardized mainstream perspectives and approaches.

The "internal logics" of gender equality in the Liberian context

Our data analysis revealed five salient themes with respect to how Liberian women conceptualize gender equality at specific sites and the ways in which they address what needs to be done to create a culture of equality. These themes included: a) Education and access to information, b) Leadership opportunities, c) Elimination of sexual and gender-based violence, d) Gender mainstreaming and financial support, and e) Contested notions of gender in/equality. Although these themes are discretely addressed, they also overlap to highlight the interconnected nature of how women attempt to negotiate gender equality for access to social and economic resources, visibility, and full participation in decision-making processes.

Education and access to information

Unsurprisingly, most women identified equal and improved access to formal education and information as salient steps toward gender equality. Participants believed that education has the potential to improve women's societal participation, which will subsequently advance their rights. One female politician argued that low levels of education discourage women from pursuing leadership positions that could be occupied to address policy issues from a gender perspective. In her view, although there have been slight increases in the number of girls attending basic school, more educational opportunities for working adult women must be provided through night schools in order to achieve gender equality:

> What I see is [that] women are still shying away from leadership. The good thing is they are going to school. If you look in the school, you see a lot of girls going in there. . . . So, when it comes to gender [equality], I still want to work with my women in the night schools.
>
> (IDI 3, female politician)

Access to information on equality and women's rights was mentioned by women leaders in the media and local NGOs who believe that such knowledge would empower women to make decisions and better prepared to address gender-based violence. A female media practitioner stated that:

> When we look at gender equality, we are talking about how to create a balance. How do we ensure there is fairness and equal opportunity? . . . Our work and our goal [in the media] is to make sure that all grassroots women have the information to fight off violence, to be empowered, to make informed decisions about their own world, to be able to claim rights they're entitled to
>
> (IDI 2, female media practitioner)

The country's 14-year civil war led to low enrollment of girls in schools, high dropout rates, and teen pregnancies. Today, women and their allies attempt to address these challenges through educational awareness campaigns led by both local and international organizations to promote girls' education. Similarly, organizations such as the Carter Center in Liberia, together with women journalists, have emphasized the importance of access to information about women's rights. Together, in September 2010, they advocated for the passage of the Liberian Freedom of Information Bill into law (The Carter Centre, 2012).

Leadership opportunities

Liberian women interpret gender equality to mean visibility in traditional systems that largely value male-dominated leadership practices, or to re-frame these systems to meet their own needs and leadership styles. For instance, community-informed mediation councils, known as *Palava huts*, are dominated by men (Dansu, 2016). However, after the war, women turned their efforts to establishing Peace Huts, modeled on the Palava Hut structure, in which they offer conflict resolution services to manage disputes (Lawson and Flomo, 2020). Women are also given chieftaincy titles in their community, something that was formerly deemed impossible. A research participant observed that:

> Now we have women town chiefs, now women can sit with men to discuss issues in the community. It wasn't like that before . . . It was a taboo for a

> woman to sit with a man to discuss in the community. The house we call a Peace Hut today was called a Palava Hut, and it was only men that were in the Palava Hut discussing issues and judging cases. Now we are there. . . . it's good work for us. When there is a problem in the community, we are no longer silent.
>
> (IDI 4, female county administrator)

Indeed, the majority of interview participants credit the Peace Hut system, education focused on political participation, and information about rights, with allowing them to "break silence" and to gain the confidence to express their needs. Additionally, and increasingly, women are contesting the notion that they should require male approval to participate in public life, or that they should be treated as property; these are beliefs that, in their view, impedes women's progress in Liberia.

In a study with male students in two Liberian universities aimed at documenting their views about women and sexual and gender-based violence (SGBV), many of the respondents state that "women don't have rights because they are property," and they use this to justify spousal rape and SGBV claiming that "you can't rape your property"'(Zwier, 2017, p. 195). These framings about the status of women in Liberia partly explain why women's groups have resorted to active advocacy work in the area of gender justice. This type of advocacy on multiple fronts is anchored in a long and robust history of feminist activities across the continent (Mama, 2017). Consequently, creating safe spaces for women in Peace Huts, holding perpetrators accountable for violence, and insisting on full political participation are some of the ways that women are challenging patriarchal discourses and practices toward gender equality for all Liberians.

In addition, women who work in the country's security sector articulated connections between security and gender equality. For some of the women we interviewed, the changing roles of female security officers from desk/administrative officers to high-ranking officials encapsulate the meaning of gender equality. A senior ranking female security personnel summarized this change as follows:

> Security and gender equality are the same. . . . Well, in the past when you heard the word, "security," the first gender you imagine[d] was a male. Nobody ever thought that women should be in security . . . the notion in the past was women in security should be restricted to a receptionist job. [But it is different now]. I worked as a security personnel and I grew through the rank and file until I became the Deputy Inspector-General. Now we are glad that most of the security institutions have developed gender policies. Now, security women and men see themselves as partners in the profession. We want women to sit at the decision-making table with the security men and come up with arguments and find solutions.
>
> (IDI 1, female security personnel)

This significant change in the security forces is partly attributable to the adoption and implementation of the UN Security Council Resolution 1325 that requires the equal representation of women in the security forces and in decision-making processes for peace, transitional justice, and development.

Many women envision gender equality to be one without restrictions on their capacity to pursue leadership positions. In one interview, a female politician addressed how patriarchy is organized to deprive women of leadership opportunities. In a conversation with her father just before he died, she lamented how he forbade her from entering politics. According to him, this arena was "dirty" and should be the reserve of men. However, she defied her father and is now among the few female politicians in Liberia. For this participant, gender equality must entail changing the patriarchal mindset of a segment of the Liberian population to value women and girls as capable leaders.

> Our environment is purely male dominated, so you have cultural barriers, you have traditional barriers, you have religious barriers that also press women down. How do we open that door? So that regardless of who you are, you can be what you want to be. If a girl wants to be a politician, hey, she should have that right, she shouldn't be told by her father 'oh it's not a place for girls.' How do we change the mindset?
>
> (IDI 5, female politician)

This question challenges the vast majority of African women who aspire to leadership positions. Notwithstanding that to date, there have [only] been two African women heads of state (Johnson Sirleaf in Liberia and Joyce Banda in Malawi), women are deeply critical of the many ways in which they are still largely expected to find fulfillment in marriage and motherhood. Women are critical of the different, sexualized standards to which they are held when they aspire to political office (e.g., a focus on dress and physical appearance, rather than on their skills and policy platforms), as well as of the lack of mentorship opportunities across business and government institutions, which are crucial to advancing gender equality for women.

Elimination of sexual and gender-based violence

Redefining the status of women as capable leaders in Liberia and eliminating SGBV were two interconnected themes that were mentioned by most of the women. Indeed, research participants argued that changing society's mindset should not be restricted to women holding political office but should constitute eradicating practices that promote violence against women and girls including SGBV. The impact of SGBV is not only physical but also mental and emotional, affecting women, the family, community, and the nation. As such, women insist that living in a violence-free society is the cornerstone for gender equality. As one policymaker who was in charge of the Department of Gender and Development explained:

> For me, my goal is seeing SGBV eliminated or reduced. And I feel that the issue of gender equality can never be successful when you have so much

> violence. There's nowhere in the world with so much violence against women and girls – and then you say you have gender equality? So, the first thing to achieve in gender equality is to ensure that we reduce or eliminate violence against women and girls.
>
> (IDI 9, female policymaker)

Although Liberia has a strong anti-rape law that came into effect in 2006 (Medie, 2013), it is difficult to enforce because of the lack of a robust statutory legal system; thus, victims of rape often rely on customary laws to address this crime. This is a phenomenon that is stringently opposed by women's groups and their allies and, in particular, by Liberian women who work in Peace Huts, all of whom insist that rape cases must be addressed in the formal legal system if pathways to gender equality are to be institutionalized across all sectors. Moreover, Liberians fear that the crime of rape has intensified due to COVID-19, a reality that has eroded fragile gender equality gains and protections, with implications for the sustainability of peace. Indeed, "According to the Ministry of Gender's statistics quoted by local media, in the first five months of 2020, more than 600 cases of rape were reported. [As well], there was a clear increase following the outbreak of the Coronavirus pandemic (Article 19)." This reality raises questions about establishing processes and mechanisms to advance gender equality that are resistant to major crisis, an issue not just for Liberia, but for global action in light of the disproportionately gendered impacts of the pandemic.

Gender mainstreaming and financial support

Liberian women constituted equality as gender mainstreaming and communal financial support. Women politicians and others in government interpret gender equality as advancing gender-informed policies on the floor of parliament and in their various political jurisdictions, particularly in budgetary allocation. In their view, women's involvement in politics positions them to call attention to the overall well-being of women, especially in health and the economy where they need the most support.

> When there's budgetary allotment, we look in there to ensure that it is gender friendly. Like the hospitals, we look to ensure that there's money there for the hospitals. We're looking to see that there's good money there for farming, because we find that most women in the hinterland, you know, they are farmers.
>
> (IDI 4, female politician)

This quote reflects the view of many scholars in gender and development studies that electing women can push a feminist-informed "women agenda," especially in the areas of maternal, reproductive, and child health policies (Cole, 2011). These are widely shared views that constitute a political platform for structural change, rather than a preoccupation with the essentialist notion that these issues are only relevant to women.

Furthermore, community organizing underpins gender equality within Liberian society. Women-led groups, such as the Peace Huts, support women financially through a small savings and loans system known as *susu.* The *susu* system is a "type of informal credit union where women pool their money to start or maintain small businesses or pay expenses" (Lawson and Flomo, 2020, p. 1864). A female policymaker describes the process, stating that

> under the various Peace Huts, they do this village savings and loans. This is a program that is now benefitting a whole lot of women in this country – where they put in their own resources, and then turn the resources over to help one another instead of going to the bank.
>
> (IDI 7, female policy maker)

Susu groups respond to the challenges of securing loans from the banks because large numbers of women do not hold land and other properties that can be used as collateral. The complication of this issue is further revealed in the social tendency among segments of the population to view women as property:

> When it comes to banks, there are times that they would ask for collateral before they give you whatever funds that you are requesting . . . besides, the interest rates at the bank is so high. And if you are trying to say that you want to give women a voice or bring them from that space, looking at property, we always say that property takes the face of a woman, she's already a poor woman.
>
> (IDI 7, female policymaker)

While *susu* groups cannot take the place of more robust financial supports for women through gender-friendly economic policies, UN Women (2014) has enhanced its financial capacities to allow more women in rural areas to access and benefit from this system.

Contested notions of gender in/equality

Among research participants, there were concerns about the disconnection between local and international conceptualization of gender equality. While most women acknowledged the importance of internationally focus interventions to achieving gender equality, they were quick to add that these interventions should be tailored to the contextual needs of Liberian women:

> The whole idea of gender equality should be contextualized to our own reality. I don't think it should be like a one-size fits all approach when it comes to adopting this whole concept. It's like, what would be the best practices and what would work well?

This quote encapsulates the view that while it is important to identify globally shared goals toward achieving gender equality, such as SDG 5, these goals need to be interpreted and contextualized to the gendered organization of day-to-day life in each country. Likewise, another participant summarized the issue as follows:

> So, how do you deal with a woman, who for years has been going for the bath bucket of her husband, cooking, waiting for him to eat before she can eat, everybody goes to bed and she's the last person? Who should benefit from an empowerment program? Because, I think a lot of our gender advocacy has focused on the women and not the family. At the end of the day, we leave the men out, we leave the children out. So, the woman has the education but she's just maybe [literally] sweeping away . . . the day in her house, because it's not going to work for her.
>
> (IDI 4, female county administrator)

This perspective circles back to the problematics of the autonomous, empowered woman at the center of gender equality initiatives that can hinder real progress for women and their societies unless attention is paid to the multidimensionality of their lives.

The five themes presented here are merely a "snapshot" of how Liberian women across societal sectors understand and articulate the contested meanings embedded in "gender equality" within a co-constitutive framework that includes political, economic, cultural, social, and rural/urban considerations. Importantly, these themes suggest that gender equality initiatives must apprehend the complexities embedded in the multiple expectations that frame women's realities, address the contradictions within "empowerment" discourses that anchor SDG 5, and adopt meaningful approaches that include their family members and communities.

Conclusion: the liminality of "gender equality"

While there are commonalities in their desire for its materialization, women's conceptualization of gender equality is informed by differences related to their socio-political and economic status, cultural positionality, and individual aspirations. Liberian women occupy liminal and oscillating positions framed by personal, collective, and structural realities within a nation-state whose geography transcends a physical territory to encompass the interests of local and global actors and forces. The pursuit of gender equality, captured in universally supported MDGs and SDGs, offers a glimpse of possible futures for gender justice in a post-conflict context where the traumas of war linger, infrastructure remains in disrepair, and sexual gender-based violence persist – all of which coalesce to inform what can be meant by "gender equality."

It is within these circumstances that Liberian women work on multiple fronts to mend the social, physical, and psychological harm/damage that scar the country's landscapes and its people, and within which they insist that gender equality is central to peace and security. What then do women's narratives reveal about the contradictions, complications, and the limitations of pursuing MDGs and SDGs in frameworks that only partially (if it all) capture new subjectivities, desires, and possibilities that emerge from conflicts? While conflicts may be ended by formal peace agreements, what becomes of the vestiges of war that are folded into everyday forms of structural violence? And what implications do they present for (re) thinking gender equality? While global and local actors, and especially Liberian women, may agree on a set of objectives toward gender equality and mainstreaming, there are contested understandings of the term, as well as competing paths leading to its materialization.

Inasmuch as women's narratives may reflect and dialogue with international articulations of gender equality, participants also proposed alternatives that contextualize equality ideals and goals more suitable for Liberian society. Ostensibly, the women we interviewed called for a type of gender equality that meets the unique needs of a post-conflict society focused on building a culture of peace and security on multiple fronts. Specifically, gender equality, with women at the forefront of defining what it means, must address the needs of the family as these emerge from a communal ontology. Otherwise, equating gender equality to the number of women and girls in school or occupying certain positions, while important, runs the risk of instrumentalizing and exploiting women's aspirations rather than transforming unequal relations of power.

Note

1 The country devil is the "*Poro* Master" of Liberia male secret cult. It is believed to have spiritual power capable of harming non-poro people. "According to deeply respected tradition, whenever the male 'devil' comes to the town, women, males and children who are non-members of the Poro secret society are forced to hide themselves away or else they risk facing great evils upon themselves and their families." Women in particular risk infertility if they come in contact the country devil (Geterminah, 2020).

References

African Union (2021) Gender Equality and Development. Available at: https://au.int/en/gender-equalitydevelopment#:~:text=The%20AU%20recognises%20that%20that,women%20in%20Africa's%20development%20agenda [Accessed 20 February 2021].

Antabe, R., Konkor, I., McIntosh, M., Lawson, E., Husbands, W., Wong, J., Arku, G. and Luginaah, I. (2021) 'I went in there, had a bit of an issue with those folks': Everyday challenges of heterosexual African, Caribbean and Black (ACB) men in accessing HIV/AIDS services in London, Ontario. *BMC Public Health*, *21*(1), pp. 1–14.

Canadian Women's Foundation (2020) Gender Equality: Our Progress Is at Risk. Available at: https://canadianwomen.org/the-facts/ [Accessed 15 November 2020].

The Carter Centre (2012) Access to Information Bill. Available at: www.cartercenter.org/peace/ati/ati-in-liberia.html [Accessed 1 February 2021].

Celis, K., Childs, S., Kantola, J. and Krook, M.L. (2008) Rethinking women's substantive representation. *Representation*, *44*(2), pp. 99–110.

Cole, S. (2011) Increasing women's political participation in Liberia: Challenges and potential lessons from India, Rwanda and South Africa. *International Foundation for Electoral Systems*, Washington.

Dansu, K. (2016) Mending Broken Relations after Civil War: The 'Palava Hut' and the Prospects for Lasting Peace in Liberia. Available at: https://media.africaportal.org/documents/KAIPTC-Policy-Brief-The-Palava-Hut-in-Liberia.pdf [Accessed 25 September 2020].

Duflo, E. (2012) Women empowerment and economic development. *Journal of Economic Literature*, *50*(4), pp. 1051–1079.

Executive Mansion (2018) Pres. Weah Declares Himself 'Liberia's Feminist-in-Chief': Recommits to Women's Cause. Available at: www.emansion.gov.lr/2press.php?news_id=4686&related=7&pg=sp [Accessed 30 December 2020].

Fleshman, M. (2010) Even with Peace, Liberia's Women Struggle: A Conversation with Activist Leymah Gbowee. Available at: www.un.org/africarenewal/magazine/april-2010/even-peace-liberia%E2%80%99s-women-struggle [Accessed 30 December 2020].

Geterminah, H.N. (2020) Women Want Feminist-in-Chief to Keep His Word. Available at: www.liberianobserver.com/news/women-want-feminist-in-chief-president-weah-keep-his-word/ [Accessed 27 December 2020].

Hanna, H. and Alfaro, A.L. (2012) The future of development in Liberia: Keeping women on the agenda. *Women's Policy Journal of Harvard*, *9*, p. 77.

Hulme, D. (2007) *The making of the Millennium Development Goals: Human development meets results-based management in an imperfect world* (No. 1607). GDI, The University of Manchester.

Jennings, Y.R. (2012) *The impact of gender mainstreaming on men: The case of Liberia.* Washington DC: George Mason University. Available at: https://search.proquest.com/docview/1010417604/fulltextPDF/D999D566D8FF49FDPQ/1?accountid=15115 [Accessed 25 October 2020].

Johnson, V. (2016) Women's Manifesto Encourages Increased Participation in Governance: Paulita Wie. Available at: www.liberianobserver.com/news/womens-manifesto-encourages-increased-participation-in-governance-paulita-wie/ [Accessed 20 November 2020].

Kabeer, N. (2015) Tracking the gender politics of the Millennium Development Goals: Struggles for interpretive power in the international development agenda. *Third World Quarterly*, *36*(2), pp. 377–395.

Khoja-Moolji, S.S. (2017) The making of humans and their others in and through transnational human rights advocacy: Exploring the cases of Mukhtar Mai and Malala Yousafzai. *Signs: Journal of Women in Culture and Society*, *42*(2), pp. 377–402.

Khurshid, A. (2016) Empowered to contest the terms of empowerment? Empowerment and development in a transnational women's education project. *Comparative Education Review*, *60*(4), pp. 619–643.

Lawson, E.S. and Flomo, V.K. (2020) Motherwork and gender justice in Peace Huts: A feminist view from Liberia. *Third World Quarterly*, *41*(11), pp. 1863–1880.

The Liberian National Gender Policy (2009) The Liberian National Gender Policy: Abridged Version. Available at: http://extwprlegs1.fao.org/docs/pdf/lbr167565.pdf [Accessed 10 December 2020].

Lomazzi, M., Borisch, B. and Laaser, U. (2014) The Millennium Development Goals: Experiences, achievements and what's next. *Global Health Action*, *7*(1), p. 23695.

Mama, A. (2017) The power of feminist pan-African intellect. *Feminist Africa 22 Feminists Organising: Strategy, Voice, Power, 1.*

Medie, P.A. (2013) Fighting gender-based violence: The women's movement and the enforcement of rape law in Liberia. *African Affairs*, *112*(448), pp. 377–397.

National Election Commission. (2020) 2020 Senatorial Election Results. Available at: https://necliberia.org/results/ [Accessed 25 August 2021].

Nwafor, A.I. (2019) Financing Post-2015 Development Goals: Shaping a New Policy Framework for Aid in Liberia. Available at: https://scholarworks.waldenu.edu/cgi/viewcontent.cgi?article=8131&context=dissertations [Accessed 27 February 2021].

Pailey, R.N. and Williams, K.R. (2017) "Is Liberia's Sirleaf really standing up for women?# LiberiaDecides." *Africa at LSE*. Available at: http://eprints.lse.ac.uk/84997/1/africaatlse-2017-09-06-is-liberias-sirleaf-really-standing-up-for.pdf

Shettima, K. (2016) Achieving the sustainable development goals in Africa: Call for a paradigm shift. *African Journal of Reproductive Health*, *20*(3), pp. 19–21.

Sustainable Development Goals (2021) Goal 5: Achieve Gender Equality and Empower All Women and Girls. Available at: www.un.org/sustainabledevelopment/gender-equality/ [Accessed 2 March 2021].

Tulay-Solanke, N. (2018) Where Are the Women in George Weah's Liberia. Available at: http://worldpolicy.org/2018/05/30/where-are-the-women-in-george-weahs-liberia/ [Accessed 20 October 2020].

UNESCO (2017) UNESCO and Gender Equality in Sub-Saharan Africa: Innovative Programmes, Visible Results. Available at: www.unesco.org/new/fileadmin/MULTIMEDIA/HQ/AFR/images/3781_15_E_web.pdf [Accessed 10 January 2021].

United Nations MDG Report. (2015) The Millennium Development Goals Report 2015. United Nations, New York. Available at: www.un.org/millenniumgoals/2015_MDG_Report/pdf/MDG%202015%20rev%20(July%201).pdf [Accessed 20 December 2020].

UN Women (2014) Liberian Women Prosper with Newfound Skills. Available at: www.unwomen.org/en/news/stories/2014/1/liberia-economic-empowerment-for-peace [Accessed 29 December 2020].

UN Women (2021) Concepts and Definitions. Available at: www.un.org/womenwatch/osagi/conceptsandefinitions.htm [Accessed 25 February 2021].

Women's NGO Secretariat of Liberia (2020) What We Do. Available at: http://wongosol.com/ [Accessed 4 January 2021].

Zwier, P.J. (2017) Human Rights for Women in Liberia (and West Africa): Integrating Formal and Informal Rule of Law Reforms through the Carter Center's Community Justice Advisor Project. *Law and Development Review*, *10*(2), pp. 187–235.

4 Global trends in women's employment in renewable energy

Continuities, disruptions, and contradictions

Rabia Ferroukhi, Celia García-Baños López, and Bipasha Baruah

Introduction

As evidenced by its inclusion as a standalone Sustainable Development Goal (SDG 5), gender equality is not only a fundamental human right but also a necessary foundation for a peaceful, prosperous, and sustainable world. There has been significant progress made globally to achieve gender equality over the last few decades: in education, employment, political participation, leadership, and legal reform. When the SDGs were framed in 2015, ensuring access to affordable, reliable, sustainable, and modern energy for all was also deemed an objective important enough to warrant a standalone Sustainable Development Goal (SDG 7). Achieving SDG 7 can, in turn, contribute toward meeting SDG 5 and other SDGs aimed at alleviating poverty (SDG 1), improving health and well-being (SDG 3), promoting quality education (SDG 4), improving access to decent work (SDG 8), and building sustainable communities (SDG 11) (IEA et al., 2020).

The International Renewable Energy Agency (IRENA) is an intergovernmental organization founded in 2009 to support countries in their transition to a sustainable energy future, and to serve as the principal platform for international cooperation, a center of excellence, and a repository of policy, technology, resource, and financial knowledge on renewable energy (RE hereafter). IRENA promotes the widespread adoption and sustainable use of all forms of RE, including bioenergy, geothermal, hydropower, ocean, solar, and wind energy in the pursuit of sustainable development, energy access, energy security, and low-carbon economic growth and prosperity. Membership in IRENA is open to states that are members of the United Nations, and to regional intergovernmental economic-integration organizations. At present, IRENA has 162 member states, with another 21 states in various stages of accession. IRENA encourages governments to adopt enabling policies for RE investments, provides practical tools and policy advice to accelerate RE deployment, and facilitates knowledge sharing and technology transfer to provide clean, sustainable energy for the world's growing population.

One of the duties IRENA has performed since its inception is to produce an annual global review of employment in the RE sector. In 2020, the review estimated that the number of jobs in the RE sector could increase from 11.5 million in

DOI: 10.4324/9780367743918-5

2019 to nearly 42 million in 2050, provided appropriate policies are put in place (IRENA, 2020a, 2020b). To enable a sustainable and equitable global transition to renewables, the industry must engage the skills and abilities of women since they constitute half of the human population. Globally, women currently represent 32% of the RE workforce – a much higher share than women in the conventional oil and gas sector (22%), but well below the 48% in the overall economy (IRENA, 2019). As in many other sectors, women are particularly underrepresented in the RE sector in jobs that require STEM training (28%) compared to non-STEM technical jobs (35%) and administrative positions (45%) (ibid.). A later survey conducted by IRENA, with responses from 920 individuals and organizations, found that women's talents and insights are even more under-utilized in the wind energy sector, where women represent only 21% of the workforce (IRENA, 2020c). The global shift to renewables is creating demand for a growing array of technical, business, administrative, economic, and legal skills, and acute skill and labor shortages are being reported globally in the RE sector (Baruah and Gaudet, 2018; IRENA, 2020c). Widening the RE talent pool is thus an instrumental reason for improving the participation of women, in addition to intrinsic reasons of gender equity and fairness. Despite clear evidence of the importance of both SDG 5 and SDG 7, as well as the interdependencies and synergies between them, empirical data and detailed information on the status and trends related to gender equality in the RE sector remain sparse.

In this chapter, we share findings from an online survey carried out by IRENA in 2018 to collect empirical data to understand global trends, opportunities, and constraints for women's employment in RE. The survey analyzes the status of women's employment in the RE sector in two distinct deployment settings – the modern context (in which universal energy access already exists but renewables are displacing or complementing fossil fuels) and the energy access context (areas where access to modern energy services, including electricity and clean cooking fuels, is presently limited or does not exist, but characterized by efforts to introduce and expand access to modern energy services). With some exceptions in Eastern and Central Europe, where energy access is still a challenge for certain segments of the population (Clancy and Feenstra, 2019), the energy context in most industrialized Organisation for Economic Cooperation and Development (OECD) member countries can be considered modern. The energy context in emerging economies and developing countries in Asia, Africa, Latin America, and the Caribbean is often much more complicated and tends to include both modern and access contexts within the same countries. The survey was designed to glean information about modern and access contexts because of the distinct obstacles due to the different nature of the challenges that women face in each of the contexts.

The survey elicited a total of 1,440 responses from 144 countries (see Figure 4.1 for geographical distribution of survey respondents).

Respondents included 1,155 individuals (808 women and 347 men) working in highly specialized technical roles, as well as in policy, legal and commercial functions, and from 285 organizations (private companies, government agencies,

Figure 4.1 Geographical distribution of survey respondents
Source: IRENA, 2019

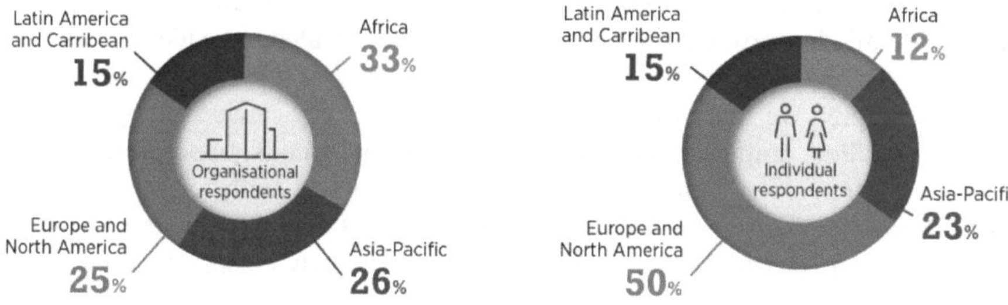

Figure 4.2 Distribution of survey respondents by region
Source: IRENA, 2019

non-governmental organizations, academic institutions, and other entities) working in the RE sector (see Figure 4.2 for distribution of survey respondents by region).

These responses provided estimates about the total share of women in the RE workforce as well as their representation in managerial and decision-making positions. The individual and institutional responses generated by the global survey also provided quantitative and qualitative insights about barriers and opportunities for women's entry, retention, and advancement in the RE sector in both modern and energy access contexts. The findings from the 2018 IRENA survey currently constitute the largest global data set on women's employment in RE.

In order to provide conceptual anchors for this research, as well as a framework of analysis for understanding survey responses, we reviewed the existing scholarly and practitioner literature on women's employment in the energy sector (both fossil fuel-based and renewable). We also reviewed research on topics such as women and technical occupations, women and part-time work, gender and care-giving, and gender and institutional sexism, which were identified in the existing literature on women's employment in the energy sector as important for understanding the broader underpinnings of women's underrepresentation in energy sector employment. We selectively present some of this literature in this chapter in order to frame and contextualize our findings about global patterns of women's employment in RE. This chapter focuses specifically on sharing findings from IRENA's 2018 survey, although we may at various points also share findings and perspectives from our previous research projects and publications. We provide references to other relevant research throughout the chapter so that those interested in learning more about gender issues in energy employment can also seek out these other resources. We hope that the issues identified by this research will provide the grounding and detail against which other related issues and research, perhaps using very different methodologies as well as broader conceptualizations of gender equality (including racial and ethnic identities, workers with disabilities, and gender-diverse LGBTQ persons) and intersectional gender analysis, can be tested, verified, and advanced.

Survey details and limitations

The survey was conducted online from October 8, 2018, to November 25, 2018. To reach a broad audience and generate a sufficiently large sample of participants, it was advertised widely through various distribution channels of IRENA and its partners, including mailing lists, newsletters, online fora and news sites, social media, e-mails from staff, and messages at national and international RE events.

The primary objective of the survey was to gather quantitative and qualitative information about women's formal employment in the RE sector, the challenges and opportunities faced by women in the sector, and suggestions to improve gender diversity. Respondents could complete the survey either as individuals or on behalf of their employers as organizations. It was an open survey, that is, anyone who could access the online link would be able to complete it.

From individuals, information was collected about their perceptions of the main barriers and opportunities to attracting and retaining women in the RE workforce, as well as suggestions for potential solutions to increase women's participation.

From representatives of organizations (e.g., human resources staff with knowledge of relevant organizational statistics), the survey asked for quantitative information about the gender distribution in the organization's workforce and the policies and measures used to support greater gender diversity.

Survey questions also distinguished between the modern energy and energy access contexts and asked respondents to identify which context they worked in. The survey was made available in five languages, namely, Arabic, Chinese, English, French, and Spanish. Participation in the survey was worldwide – with respondents from 144 countries. The response rate from China was generally low, even though the country is a major player in the Asian and global RE sectors. Responses from organizations were quite evenly distributed across the main

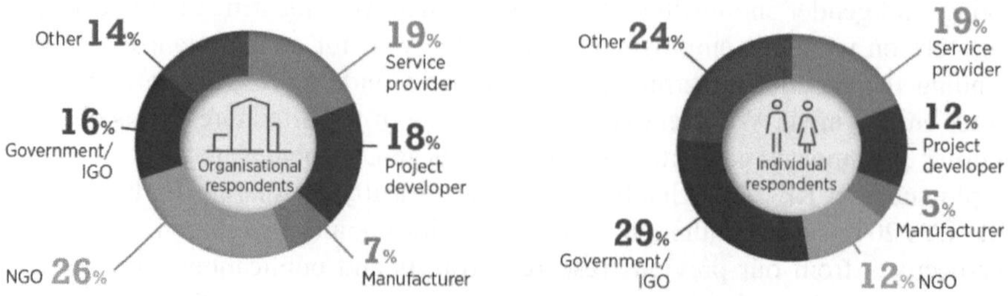

Figure 4.3 Distribution of survey respondents by type of organization
Source: IRENA, 2019

regions of the world, with many responses from organizations located in Africa. By contrast, half of the responses from individuals came from Europe and North America.

With respect to the organizations for which respondents work, the survey generated fewer responses from private sector organizations than from governmental, inter-governmental, and non-governmental organizations (see Figure 4.3).

In emerging economies and developing countries, the informal sector is a major driver of the economy. For example, in India, 88.2% of the employed population are informal workers, 82.7% in Kenya and 92.9% in Nigeria (ILO, 2018). Women often constitute the majority of informal sector workers in these countries (ibid). Since the IRENA survey was designed to study formal employment in the RE sector, it did not provide specific insights into challenges faced by women in informal RE activities, such as fuel collection and biomass production and processing. Efforts made by other researchers and intergovernmental organizations to provide estimates of informal employment in RE have been documented elsewhere (Baruah, 2015, 2017) and suggest that millions of women are engaged informally in RE. Future research aimed at collecting empirical data on informal employment in RE is crucial.

Major findings and discussion

The IRENA survey revealed that women represent 32% of the full-time employees of responding RE organizations, which is substantially higher than the 22% average in the global oil and gas industry (see Figure 4.4).

The share of women in the worldwide oil and gas workforce is much lower than in manufacturing, finance, education, health, and social work, and lower than the average in the overall global workforce (Rick et al., 2017). While women hold 27% of entry-level jobs in the oil-and-gas sector that require a college degree, and 25% of midcareer-level jobs, their share is only 17% in senior and executive roles. Only one in a hundred CEOs in the fossil fuel sector is a woman (ibid.). As a young and more dynamic sector, RE may be open to diversity and change in ways that are harder to effect in the older and more mature fossil fuel sector.

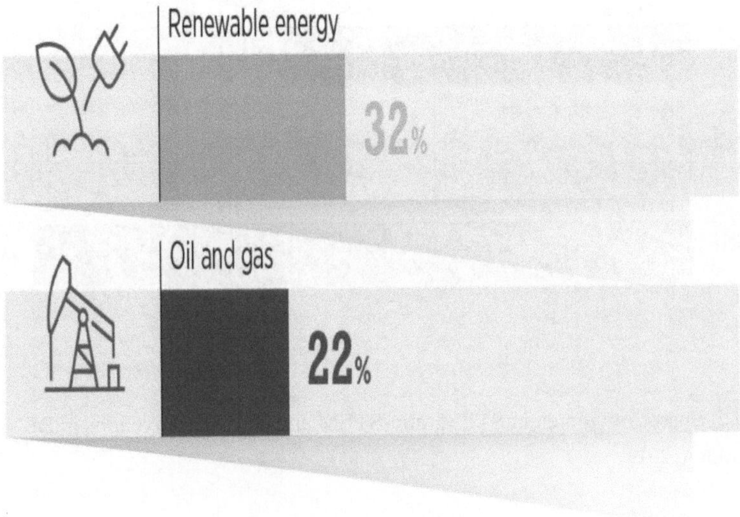

Figure 4.4 Share of women in RE and Fossil Fuels
Source: IRENA, 2019

This is borne out by findings from an earlier survey carried out by IRENA in 2016 of 90 RE companies in 40 countries, which revealed that women held 28% of technical positions and 32% of senior management positions in the RE industry. However, the findings from the 2018 IRENA survey reveal that although women are better represented at all levels of the RE sector than they are in the oil-and-gas sector, women continue to be underrepresented in the RE sector in jobs that require science, technology, engineering, and math (STEM) training (28%) compared to non-STEM technical jobs (35%) and administrative positions (45%) (see Figure 4.5).

Although renewables employ more women than fossil fuels, our findings suggest that women face persistent barriers to entry, retention, and to advance to senior executive and leadership positions. Removing these barriers is essential to meet the growing demand for skills in an expanding RE industry.

The fact that women face specific barriers to retention and advancement in the RE sector is also evident from the finding that male and female respondents to the 2018 IRENA survey had similar levels of education. Almost three-quarters of respondents (71%) reported having a university degree in a STEM subject, with most of the rest holding postsecondary degrees in non-STEM subjects, and 4% having high-school or diploma qualifications. The highest degree attained by respondents was a master's degree.

As mentioned previously, the survey was designed to shed light on both the modern energy and energy access realms. A small number of individual (11%)

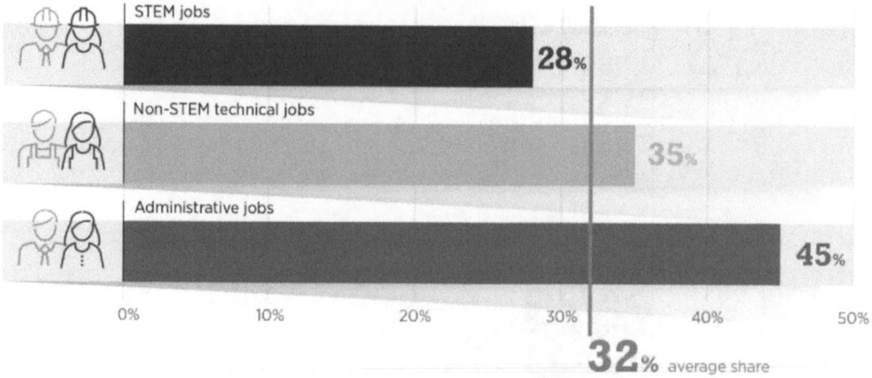

Figure 4.5 Shares of women in STEM, non-STEM and administrative jobs in RE
Source: IRENA, 2019

and organizational respondents (16%) indicated that they were working exclusively in the access context. A larger share – a quarter of all participating organizations and almost half of all individual participants – operated exclusively in the modern context. The largest share of respondents (57% of organizational respondents and 42% of individual respondents) indicated working in both modern and access contexts.

Social perception of gender roles

Regardless of whether they worked in the modern or energy access contexts, respondents identified societal perception of gender roles, cultural and social norms surrounding women's employment, and male-biased hiring practices in the energy sector as the top three barriers for women's entry into the RE sector (see Figure 4.6). These findings resonate with other recent research conducted to understand women's underrepresentation in the RE sector in various industrialized, emerging, and developing economy contexts. For example, Baruah (2017) emphasizes how societal misperceptions of women's incompetence in technical occupations present an impediment for women's optimal participation in RE. Women in technical occupations are deemed less competent than men, even when they have more education and experience than their male counterparts (ibid).

Barriers to advancement

When asked to identify barriers to advancement in the RE sector, survey respondents unanimously emphasized the existence of a glass ceiling (i.e., unacknowledged or invisible barriers to advancement) in the RE sector. They cited cultural and social norms, lack of flexibility in the workplace, and lack of mentors as the

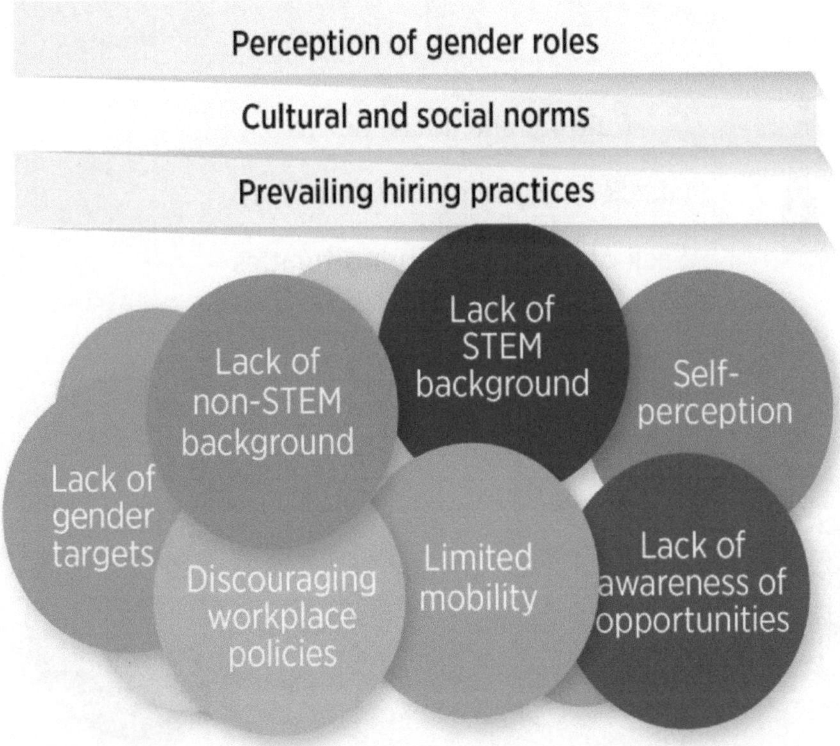

Figure 4.6 Barriers to entry for women in the RE sector
Source: IRENA, 2019

top three barriers for women's advancement in careers in RE (Figure 4.7). The barriers to advancement revealed by this survey have been emphasized elsewhere. For example, in a study about women, gender equality, and the energy transition in the European Union, Clancy and Feenstra (2019) identify male-biased hiring and promotion practices and lack of flexibility in work hours as major barriers to women's advancement in the RE sector. Also, as noted by Baruah (2019), women encounter both "sticky floors and glass ceilings" in the RE sector. In other words, careers may never get off the ground because of persistent and confining stereotypes of feminized roles. And the absence of role models and gender-balanced initiatives makes moving up the ranks more challenging for women (ibid).

Barriers to participation by geography

The regional distribution of perceived barriers to women's participation in the RE sector provided interesting results. Although cultural and social norms were

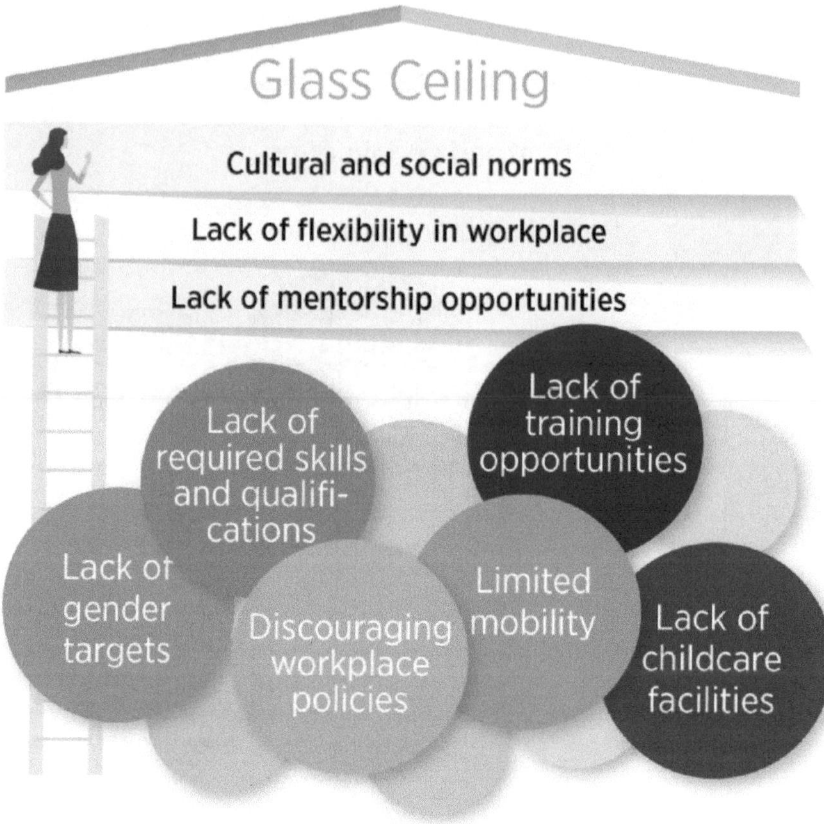

Figure 4.7 Barriers to women's advancement in the RE sector
Source: IRENA, 2019

identified as barriers for women's participation in the RE sector by respondents in all geographical settings, respondents from Europe and North America identified them as a bigger barrier than respondents from Latin America and the Caribbean, Asia, and Africa (see Figure 4.8). Respondents from Europe, North America, and the Asia-Pacific regions also identified unequal asset ownership as bigger barriers than respondents from Africa, Latin America, and the Caribbean. Respondents from Africa and the Asia-Pacific region identified lack of skills as the biggest barrier to women's entry into the RE sector. These findings are unique to the 2018 IRENA survey. They have not been written about elsewhere in the existing literature on women's employment in the energy sector. Of course, the fact that respondents from countries in Africa, Asia, Latin America, and the Caribbean did not identify cultural and social norms as the biggest barrier does not mean that such barriers do not exist in these contexts. It may just speak to the fact that lack of

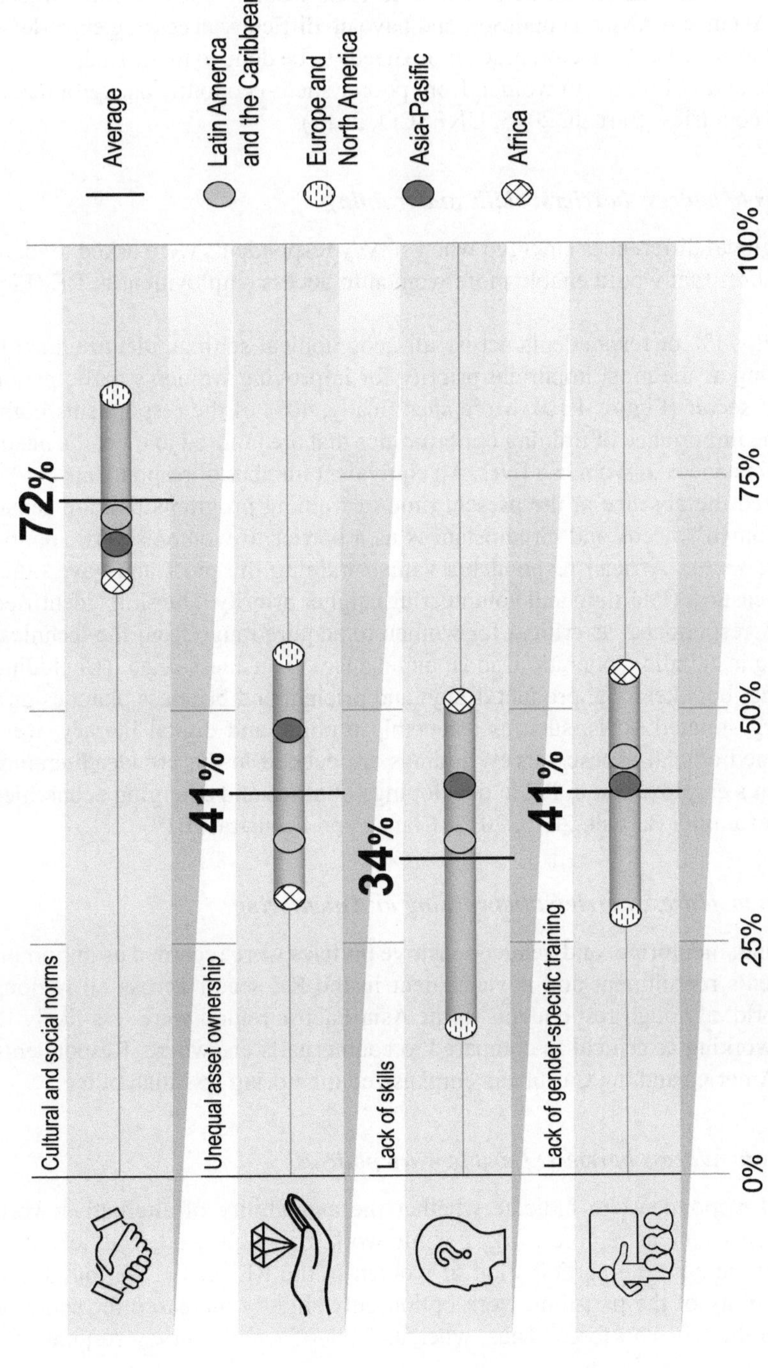

Figure 4.8 Regional distribution of barriers to women's employment in RE

Source: IRENA, 2019

access to training and skills is perceived as a more immediate and practical barrier than cultural and social norms. Indeed, the existing research on the topic emphasizes that although large numbers of middle-class women study STEM subjects in many Asian and African countries, and have no difficulty accessing entry-level technical jobs in the RE sector, much more needs to be done to make such training accessible and affordable to women from poorer socio-economic backgrounds in the same countries (Baruah, 2015; UNESCO, 2015).

Measures to address barriers: skills and training

Other regional differences emerged when survey respondents were asked to identify measures that would enable more women to access employment in RE (Figure 4.9).

Overall, 71% of respondents across all geographical settings identified skills and training as the most important priority for improving women's participation in the RE sector (Figure 4.10). More specifically, 40% of the respondents highlighted the importance of training opportunities that are tailored to meet the needs and circumstances of women's lives. An equivalent number of respondents (41%) emphasized the absence at the present time of training programs that are sensitive to women's needs and circumstances as a barrier to women's participation in the RE sector. African respondents valued training the most and gave other related measures (interning and volunteering) higher priority. The skills identified by survey respondents as critical for women to acquire range from the technical (involving installation, operation, and maintenance) to business-related (including accounting, bookkeeping, product design and pricing, and business plan design). Non-energy-related skills, such as leadership training and digital literacy, were also deemed crucial. These survey findings are echoed in the broader literature on women's employment in RE in developing countries and emerging economies (see, for example, Baruah, 2015, 2017; Clancy and Feenstra, 2019).

Measures to address barriers: networking and mentoring

Networking, mentoring, and gender-sensitive policies were regarded as important for women's recruitment and advancement in the RE sector across all regions of the world, although respondents in the Asia-Pacific region were less likely to view networking as critical as compared to counterparts elsewhere. Respondents in Latin America and the Caribbean emphasized mentoring less than others.

Measures to address barriers: flexible work policies

We asked respondents to indicate whether the availability of alternative work arrangements such as remote work, flexible work, part-time work, and job sharing would increase the participation of women in the RE sector. We found that the availability of the part-time work option only slightly increased the share of women in the RE workforce. Thus, whereas 32% of female survey respondents

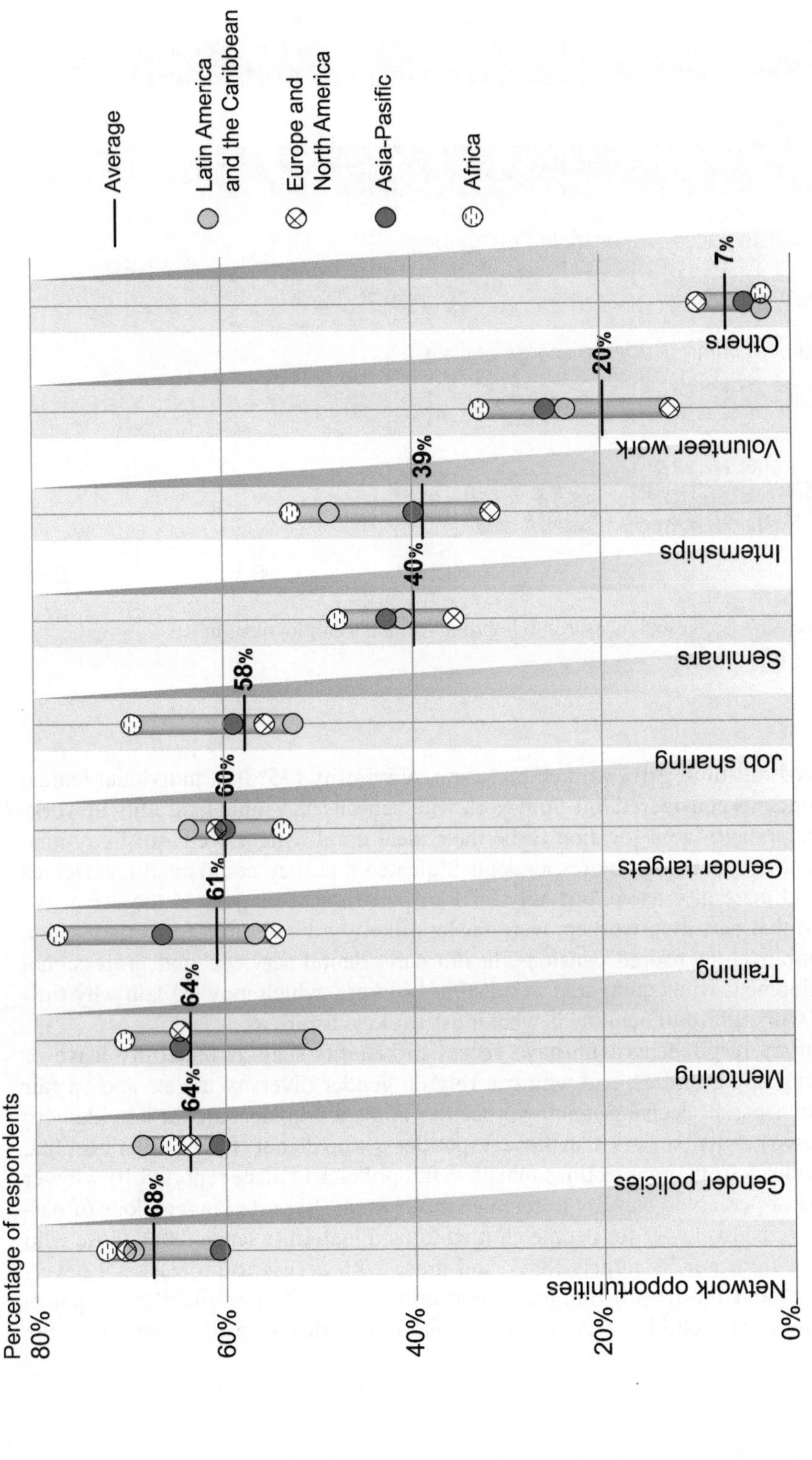

Figure 4.9 Suggested measures to enable more women to access employment in RE

Source: IRENA, 2019

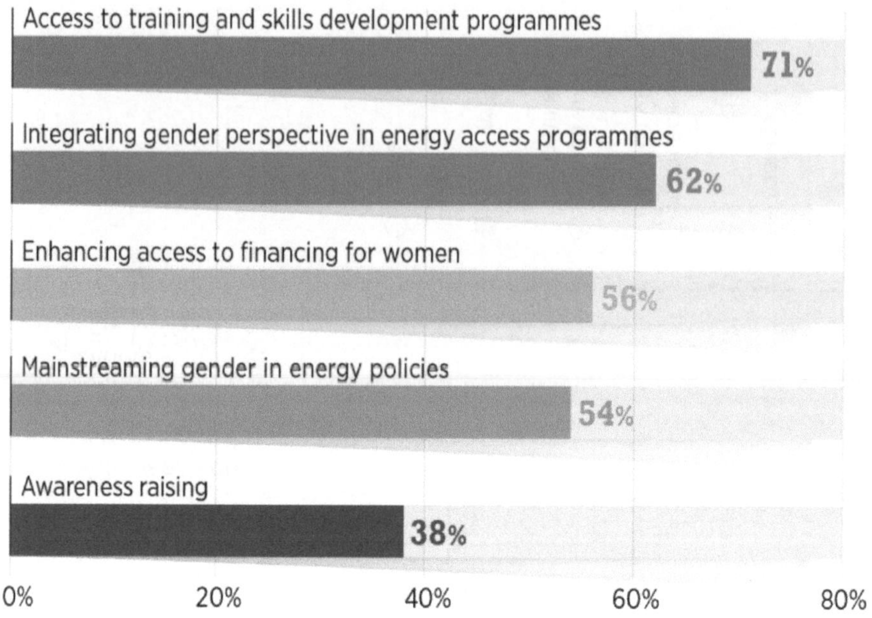

Figure 4.10 Suggested measures to improve women's employment in RE
Source: IRENA, 2019

worked full-time, 36% worked part-time. A majority (75%) of individual female respondents considered full-time work with benefits and some flexibility in working hours and work location to be their ideal employment scenario. Sixty-nine percent of female survey respondents indicated that they could avail themselves of paid maternity leave, but only 34% of part-time workers said the same. We found that part-time workers were far less likely to be eligible for benefits such as employer-sponsored pensions, health care, dental services, and professional development when compared to full-time workers, which may explain why full-time work with full benefits is what most workers aspire to.

Survey respondents who have access to benefits such as maternity leave or training opportunities, and who can rely on gender diversity targets and on fair and transparent decision-making processes in the workplace, are far less likely to perceive gender barriers than those respondents who do not enjoy similar benefits. Individuals from organizations that have fair policies in place reported 10% lower levels of perceived barriers to female employment. Reported perceptions of barriers were also lower for people entitled to paid maternity leave (9%), those who benefit from gender targets (8%), and those with access to professional development and on-the-job training opportunities (7%). The availability of paternity leave and childcare facilities also had a significant positive impact on the perception of barriers to female employment, although relatively few employers offered these benefits.

Awareness of gender issues

One of the most insightful findings from the survey was the gender difference in the response rates and the perception of barriers in the RE sector. Both men and women were invited to respond to the 2018 IRENA survey, but almost 70% of respondents were women. This may serve as an indication that awareness of gender issues in the RE sector is itself driven significantly by gender. Along the same lines, although 75% of female respondents identified gender as a barrier in the RE sector, only 40% of male respondents did so. Similar gender differences emerge regarding perceptions of pay equity in the RE sector. Sixty percent of male respondents and only 29% of female respondents agreed with the statement that women and men were paid equally for the same work in the RE sector. Most men working in the RE sector, presumably including those with responsibilities for making policy decisions, are unaware that women face specific barriers entering the sector and advancing within it. This is borne out by findings from the survey about the gender composition of boards of directors of organizations in the RE sector. Women are underrepresented on boards of private sector companies, governmental agencies, and intergovernmental organizations (IGOs), and NGOs that work on energy issues (see Figure 4.11). Women are most poorly represented on the boards of private sector companies, government agencies, and IGOs. Men represent at least 75% of board members for private sector organizations that responded to the survey. One survey respondent summed up why gender equality does not receive much attention within the RE sector succinctly: "If you don't even know you have a problem, you're certainly not going to fix it."

Benefits of gender diversity in the workplace

Research conducted all over the world confirms that gender diversity in organizations, and wage equity between women and men, results in higher profits and better returns on investment. In its study of almost 22,000 firms across the globe, the Peterson Institute for International Economics discovered that a company with 30% women leaders can add up to 6% points to its net margin, compared to other companies in the same industry. Across the economy, the percentage of women corporate officers is positively linked to better financial performance (Noland et al., 2016). Another study found that companies with more women board members, on average, outperform those with fewer women by 53% on return on investment, 42% on return on sales, and 66% on return on invested capital (Joy, 2008). Similar findings have emerged for women in executive positions – companies with higher percentages of women decision makers financially outperform their industry peers. Across all sectors of the economy, the percentage of women corporate officers is positively linked to better financial performance. A 30% critical mass of women as executive officers and board members has

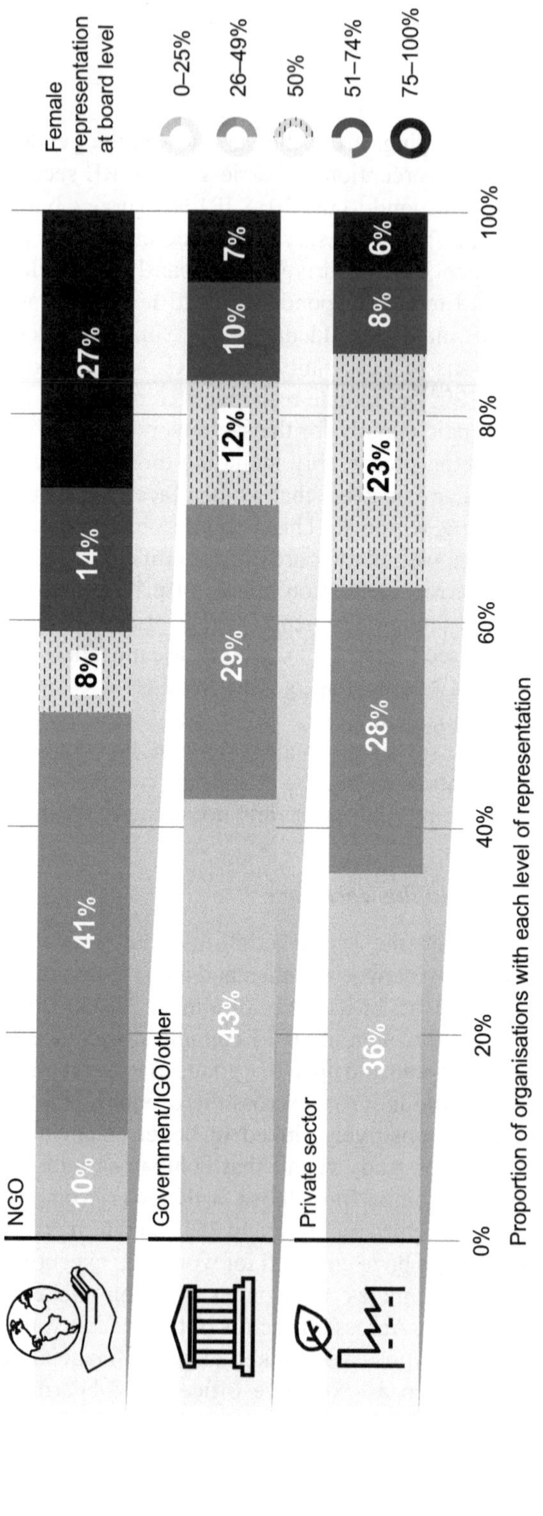

Figure 4.11 Gender composition of boards of directors in the RE sector

Source: IRENA, 2019

been demonstrated to have the most positive impact on company performance (Wagner, 2011). Gender balance in male-dominated professions contributes to the improvement in working conditions for both men and women, with positive effects on well-being, work culture, and productivity (WISE, 2017).

The business case – stronger economic outcomes if more women are employed – is often the most powerful rationale for more companies to institute gender diversity and wage equity policies (Baruah, 2018). Therefore, collecting gender-disaggregated data on employment in the RE sector and documenting and publicizing the economic benefits of gender diversity and inclusion is a useful strategy for convincing RE companies to recruit, retain, and promote more women.

The broad lesson that emerged from IRENA's 2018 online survey is that while progress has been made in building gender diversity and inclusion in RE employment, much remains to be done to ensure that women derive optimal benefits from employment in the sector, and concurrently that the RE sector is able to draw upon a larger workforce than it currently benefits from.

Conclusions and policy recommendations

The global transition to RE is creating many benefits for the economy and the environment, including new sources of employment. Women's contributions – their talents, skills, and perspectives – are critically important in supporting this growing industry during the global transition toward a more sustainable energy system that could benefit all of humanity.

The findings from IRENA's 2018 survey of 1,440 respondents working in the RE sector in 144 countries worldwide revealed that women represent 32% of full-time employees of responding organizations, and a slightly lower number (about 31%) of female respondents are in mid-level management positions. These numbers are significantly more promising than in the global oil and gas industry, where women make up an estimated 22% of the workforce and represent 25% of mid-career level jobs.

A broad variety of policies and measures can ensure that women derive equal access and benefits from participation in the RE sector and from the energy transition more broadly. At the same time, the RE sector needs to be able to draw upon a larger talent pool. Responses to the survey revealed that although women represent 45% of general administrative jobs within participating organizations, they hold a lower number (28%) of the positions that require STEM training. Boosting women's participation in STEM education and employment is an action area that the RE sector must prioritize. At the same time, non-technical career paths – of which there are many – need to be given greater visibility and stature.

In regions of the world where large segments of the population lack reliable or affordable access to modern energy services, new entrepreneurial and livelihood opportunities arise for women. Globally, 70% of the world's poorest people are women and children, which means that gender equity must be an integral part of efforts to expand modern energy access and reduce gender inequality and poverty (Pearl-Martinez and Stephens, 2016). Thus, enabling more women to engage in the sector in both the access and the modern contexts can simultaneously advance gender equality and empowerment objectives as well as the RE sector's need for skills.

Several policy recommendations and action areas emerge from the survey and the existing literature. These include the need to proactively plan and implement (i.e., mainstream) gender equity in energy sector frameworks; to tailor training and skills development programs to women's specific needs; to work to both attract and retain talent in the RE sector; and to challenge cultural and social norms. These action areas point to commonalities in both the modern energy and energy access contexts, despite considerable differences between them. We elaborate briefly below on each.

Mainstreaming gender in energy sector frameworks

Almost half (49%) of the survey respondents identified the lack of gender-sensitive policies in the energy sector as a key barrier for women's employment and advancement in the RE sector. Yet, few countries have introduced meaningful policies to promote gender equity in the sector (Rojas et al., 2015).

A United States Agency for International Development (USAID) and International Network on Gender and Sustainable Energy (ENERGIA) study of RE policies found that only six of 33 examined countries (18%) included gender keywords and considerations. Conversely, national gender equity policies rarely include any targets specific to equity in access to energy services, or employment in the energy sector (Pearl-Martinez, 2014).

Energy policies and programs – regardless of whether they are driven by governments, civil society, private corporations or international aid agencies – should integrate women's experiences, expertise, capacities, and preferences, to avoid risking reinforcing the gender gap between men and women in both the modern and energy access contexts.

Gender mainstreaming is the process of analyzing legislation, regulations, taxation, and specific projects for their effect on the status of women in society (Swirski, 2002). The basic assumption of gender mainstreaming is that public policy affects men and women differentially, stemming from the different roles women and men play within their families, communities, and the economy at large. For women to enjoy equal opportunities, mainstreaming gender equity concerns and solutions into national energy sector frameworks is critical. Gender audits can be an effective instrument in identifying gender gaps across the energy sector landscape and setting a baseline for future gender mainstreaming efforts at the policy and institutional level (IRENA, 2013).

In the access context, greater efforts are needed to engage women along different segments of the off-grid RE value chain. A gender equality perspective needs to be integrated from the very beginning in the design, implementation, and monitoring of energy access programs. For example, Zambia's National Energy Policy identifies measures to mainstream gender considerations in all energy access programs and highlights the role of women not only as beneficiaries but as also active energy providers and entrepreneurs within the sector (Clancy et al., 2017; ENERGIA, 2011). This is a good example of a gender-transformative approach that views women not simply as primary end users and beneficiaries, but as actors in the design and delivery of energy solutions.

Tailoring training and skills development programs

A lack of appropriate skills and training for employment in the RE sector remains a key barrier for women in both the modern and access context. This affects women seeking entry into the RE sector, as well as those who are already employed. Low enrollment rates in STEM courses translate into continued under-representation of women in technical roles in the RE industry.

Raising awareness of career opportunities, adapting curricula and training as well as creating entry points, such as internships, co-op programs, and apprenticeships, is likely to attract more women into relevant fields. A wide range of actors can play an important role in such efforts, including governments, educational institutions, the private sector, and advocacy organizations.

In responding to the changing landscape of the energy sector, universities, colleges, and other educational institutions should make technical training programs more versatile to enable cross-sectoral transition within the energy sector. Universities should consider integrated programs (covering both renewables and non-renewables), building cross-disciplinary and connected labs, adding courses to respond to digitization trends, and offering more courses, such as business and project management, to help students increase their employability (Baruah, 2018). Non-STEM fields, such as environmental studies, public policy and administration, law, business and health, which tend to enroll large numbers of female students, are also important potential areas of recruitment (ibid.).

Professional networks and personal connections play a big role in access to career information in the RE sector. Strengthening mentoring, outreach presentations and visits, student networks, and temporary work placements can help level the playing field for women. In the energy access context, women's participation in training and skills development programs is critical. Many of the skills needed in the off-grid renewables' value chain can be developed locally. Organizations like Solar Sister, Grameen Shakti, Barefoot College, Hivos, ENERGIA, and the Self-Employed Women's Association (SEWA) have demonstrated that customized solutions for training and opportunities for cross mentorship can substantially increase women's participation in the sector (Baruah, 2015).

Attracting and retaining talent

Women want to work in the RE sector for similar reasons as men: decent incomes, good benefits, company reputation, availability of work, and opportunities to build careers. Yet, female employees often face the double burden of work and family responsibilities, making it difficult for them to remain employed and advance at par with their male counterparts in careers in RE. The survey revealed significant differences among employers in accommodating employees' caregiving needs and multiple responsibilities.

More RE companies should institute policies such as parental leave, flexible work hours, telecommuting, and working part-time. Such policies, in combination with gender equity in wages, support for childcare, and equal opportunities for professional advancement, will ensure that more early and midcareer women find it worthwhile not just to remain in their jobs, but also possible to move up the professional ladder.

An important distinction must be made between women's *representation* and *participation* in the RE sector. At lower than 15% female representation, it is not uncommon for women, because of their minority status, to be made to feel marginal and "invisible" in decision-making processes. Establishing critical mass is important for creating more supportive institutional environments in which women can overcome potential reticence and speak out on issues and concerns in the presence of colleagues (Agarwal, 2010). Toward this end, RE employers should set gender diversity targets in junior, mid-level, and senior positions in all occupations in which they are currently underrepresented, namely, trades, production, and technical and management positions.

Challenging cultural and social norms

Prevalent cultural and social norms strongly influence the success of gender goals in the RE sector. Concrete actions must be taken to reduce barriers to entry for women, establish an enabling environment for retention and advancement, and level the playing field through equal pay and workplace policies. Strengthening the visibility of the diverse roles women are already playing in the energy sector may be one way to challenge entrenched social and cultural norms.

In the access context, there are examples emerging from countries such as India and Indonesia about renewables enabling some women not just to forge a path out of poverty but also to become agents of social and economic transformation in their communities. The "solar mama" program at the Barefoot College in India is a well-documented case study of the democratizing power of off-grid RE solutions and the transformative potential of training women in rural areas. The program has trained over 1,000 women from more than 80 countries, leading to the deployment of at least 18,000 solar systems. The trainees are often nonliterate or semi-literate women who maintain strong roots in their rural villages and have the potential to play a key role in bringing off-grid solar solutions to remote, inaccessible villages. The initiative works to demystify the technology and place it in the hands of local communities. Over a period of six months, trainees receive instruction on assembly, installation, operation, and maintenance of solar lanterns, lamps, parabolic cookers, water heaters, and other devices. The women return to their villages with equipment to deliver sustainable electricity to their community and become mentors for other women in their communities (Deshpande, 2017). Participating in the RE sector has provided some women with a meaningful platform for questioning and subverting oppressive social norms and practices such as dowry, child marriage, and domestic violence (IRENA, 2018). In Indonesia, the Wonder Women program, an initiative of the non-profit Kopernik, trains women to expand last-mile electricity access through off-grid solutions. Since 2013, the program has recruited more than 500 "wonder women," who have sold more than 55,000 clean energy technologies (e.g., solar lighting solutions) reaching more than 250,000 people in some of Indonesia's poorest and most remote areas. The program provides training to female social entrepreneurs to develop their capacity to build and sustain businesses. The training focuses on technology use and maintenance, sales and marketing, bookkeeping and financial management, and public speaking. The entrepreneurs sell from home, through their networks, at market stalls and small shops, or at community events. A survey conducted after 12 months of program implementation found that 21% of participants felt more empowered within their families, taking on a greater role in household decision-making. Almost half of the survey's respondents perceived an improvement in their status and 19% felt more empowered within the community. Wonder women often become a pillar of support and inspiration for other women in the village, encouraging them to join the program or take up other economic activities (IRENA, 2018). As gender-sensitive training, education, apprenticeships, employment placement, and financial tools are adopted more widely, more women may be able to step into such roles and, in turn, contribute directly to the accomplishment of SDG 5 and SDG 7, and indirectly to the other Sustainable Development Goals.

Future research

The scarcity of gender-disaggregated empirical data is a major barrier in the effort to enhance awareness of the challenges and to improve the gender balance in RE employment. Without data, there is no visibility. And without visibility, there is no policy priority. Efforts to improve quantitative and qualitative data gathering are thus essential. This is true for both the modern energy and energy access contexts and across all regions.

Although organizational responses to IRENA's 2018 survey were evenly distributed across the main regions of the world, half of the individual responses came from Europe and North America. Also, the survey generated fewer responses from private sector organizations than from governmental, inter-governmental, and non-governmental organizations. Future data collection efforts should strive for a more comprehensive representation of the RE sector from private sector organizations, utilities, manufacturers, and large companies operating under power purchase agreements. They should also strive for better geographical coverage, especially from countries such as India and China that are key players in the global RE industry.

Government statistics will generally need to capture employment in the sector much better than is the case today, building gender disaggregation into these efforts from the beginning. A wide range of actors, including academic and non-academic researchers, advocacy groups, professional associations, international organizations, NGOs, policy institutes, and think-tanks, can contribute to building a gender-disaggregated evidence base in RE.

Context matters greatly for understanding both gender barriers and solutions; this requires more detailed examination of gender dimensions in different regions and countries, for different types of RE technologies, and for different scales of deployment. For instance, gender equity issues in large-scale grid-connected RE projects (utility-scale solar, wind, geothermal, or hydropower, for example) have not yet been researched extensively. Future research should identify guidelines and strategies in this context. Off-grid RE initiatives have generated significant new economic opportunities for women in the access context. However, available evidence of successes, remaining challenges, and long-term sustainability of such initiatives is currently largely anecdotal.

A better understanding is also required of wider social and economic policies that are necessary to optimize livelihood initiatives in the renewables sector. In particular, the creation of permanent and stable sources of income often remains a challenge for women who have been trained to build, install, repair, and sell solar systems, improved cook stoves, or other RE solutions. More women can gain optimal traction from RE initiatives if there are other supportive gender-sensitive social and economic policies. Since women's ability to take advantage of new RE-related employment options is, to begin with, often constrained by legal or social barriers that limit their education, property rights, land tenure, and access to credit, it is crucial that social and economic policies go beyond energy sector planning to enhance economic opportunities for women (Baruah, 2015). Therefore, in both the access and modern energy contexts, analytical efforts and policy initiatives that go beyond the confines of the energy sector itself may become increasingly necessary to address the gender dimension adequately.

There have been significant advancements globally in expanding and strengthening social protection policies in recent years, as more countries transition toward developing welfare systems. Some strategies that are being tried in African, Asian, and Latin American countries include basic income schemes, as well as conditional and unconditional cash transfer programs that enable poor women to make priority decisions for themselves and their dependents. Programs like Brazil's Bolsa Familia, Mexico's Prospera, Mali's Social Cash Transfer initiative, and India's basic income pilot are hopeful developments given that structural inequality constrains an individual's ability to exercise rights and demand entitlements, such as decent employment (Campello and Neri, 2014; Davala et al., 2015; Mary Robinson Foundation, 2016). In the interest of enhancing the economic benefits and social outcomes of growing employment opportunities, the RE sector may find it necessary in the future to engage in research and policy aimed at understanding how to strengthen and expand social protection infrastructure in both the modern and energy access contexts in the pursuit of a just and equitable energy transition. Strengthening social protection policies around the world will lead to the advancement and optimization of both SDG 5 and SDG 7, and the synergies and interdependencies between the two goals.

There is growing recognition that universal energy access is unlikely to be achieved without addressing the need for gender equality (IEA, 2019; IEA et al.,

2020). Future research and evidence building at the intersections of SDG 5 and SDG 7 is therefore critical.

References

Agarwal, B. (2010) 'Does Women's Proportional Strength Affect Their Participation? Governing Local Forests in South Asia', *World Development*, 38(1), pp. 98–112.

Baruah, B. (2015) 'Creating Opportunities for Women in the Renewable Energy Sector: Findings from India', *Feminist Economics*, 21(2), pp. 53–76.

Baruah, B. (2017) 'Renewable Inequity? Women's Employment in Clean Energy in Industrialized, Emerging and Developing Economies', *Natural Resources Forum*, 41(1), pp. 18–29.

Baruah, B. (2018) *Barriers and Opportunities for Women's Employment in Natural Resources Industries in Canada*. Ottawa: Natural Resources Canada.

Baruah, B. (2019) 'Addressing the Diversity Challenge in Energy Sector Recruitment', *Modern Diplomacy*, 6 July [online]. Available at: https://moderndiplomacy.eu/2019/07/05/addressing-the-diversity-challenge-in-energy-sector-recruitment/ (Accessed: 16 November 2020).

Baruah, B. and Gaudet, C. (2018) 'Creating and Optimizing Employment Opportunities for Women in the Clean Energy Sector in Canada', *Smart Prosperity Institute*, 18 May [online]. Available at: https://institute.smartprosperity.ca/library/research/creating-and-optimizing-employment-opportunities-women-clean-energy-sector-canada (Accessed: 16 November 2020).

Campello, T. and Neri, M. (2014) *Bolsa Família Program: A Decade of Social Inclusion in Brazil*. Brasilia: Institute for Applied Economic Research.

Clancy, J., Daskalova, V., Feenstra, M., Franceschelli, N., and Sanz, M. (2017) *Gender Perspective on Access to Energy in the EU*. Brussels: European Union.

Clancy, J. and Feenstra, M. (2019) *Women, Gender Equality and the Energy Transition in the EU*. Brussels: European Union.

Davala, S., Jhabvala, R., Standing, G., and Mehta, S. (2015) *Basic Income: A Transformative Policy for India*. New Delhi: Bloomsbury.

Deshpande, V. (2017) 'Inside Barefoot College: Where Women from across the World become Solar Engineers', *The Economic Times*, 18 June [online]. Available at: https://economictimes.indiatimes.com/industry/energy/power/inside-barefoot-college-where-women-from-across-the-world-become-solar-engineers/articleshow/59195677.cms (Accessed: 3 February 2021).

ENERGIA (2011) *Zambia Gender and Energy Mainstreaming Strategy 2011–2013*. The Hague: ENERGIA.

IEA. (2019) 'Status Report on Gender Equality in the Energy Sector', *C3E International*, 17 June [online]. Available at: www.cleanenergyministerial.org/sites/default/files/2019-06/Status%20Report%20on%20Gender%20Equality%20in%20the%20Energy%20Sector_0.pdf (Accessed: 3 February 2021).

IEA, IRENA, UNSD, World Bank, and WHO. (2020) *Tracking SDG 7: The Energy Progress Report*. Washington, DC: World Bank.

ILO (2018) *Women and Men in the Informal Economy: A Statistical Picture* (third edition). Geneva: International Labour Office.

IRENA (2013) *Renewable Energy and Jobs: Annual Review*. Abu Dhabi: IRENA.

IRENA (2016) *Renewable Energy and Jobs: Annual Review*. Abu Dhabi: IRENA.

IRENA (2018) 'Indonesia's "Superheroines" Empowered with Renewables', *IRENA Articles*, 22 April [online]. Available at: http://irena.org/newsroom/articles/2018/Apr/

Indonesias-Superheroines-Empowered-with-Renewables (Accessed: 14 November 2018).

IRENA (2019) *Renewable Energy: A Gender Perspective*. Abu Dhabi: IRENA.

IRENA (2020a) *Renewable Energy and Jobs: Annual Review*. Abu Dhabi: IRENA.

IRENA (2020b) *Global Renewables Outlook: Energy Transformation 2050*. Abu Dhabi: IRENA.

IRENA (2020c) *Wind Energy: A Gender Perspective*. Abu Dhabi: IRENA.

Joy, L. (2008) 'Advancing Women Leaders: The Connection between Women Board Directors and Women Corporate Officers', *Catalyst*, 15 July [online] Available at: www.catalyst.org/research/advancing-women-leaders-the-connection-between-women-board-directors-and-women-corporate-officers/ (Accessed: 1 October 2020).

Mary Robinson Foundation (2016) *The Role of Social Protection in Ending Energy Poverty Making Zero Carbon, Zero Poverty the Climate Justice Way a Reality*. Dublin: MRF.

Noland, M., Moran, T., and Kotschwar, B. (2016) *Is Gender Diversity Profitable? Evidence from a Global Survey*. Washington, DC: Peterson Institute for International Economics.

Pearl-Martinez, R. (2014) *Women at the Forefront of the Clean Energy Future*. Washington, DC: IUCN-USAID.

Pearl-Martinez, R. and Stephens, J. (2016) 'Toward a Gender Diverse Workforce in the Renewable Energy Transition', *Sustainability: Science, Practice and Policy*, 12(1), pp. 8–15.

Rick, K., Martén, I., and Von Lonski, U. (2017) 'Untapped Reserves: Promoting Gender Balance in Oil and Gas', *Boston Consulting Group*, 12 July [online] Available at: www.bcg.com/en-ca/publications/2017/energy-environment-people-organization-untapped-reserves (Accessed: 16 November 2020).

Rojas, A., Prebble, M., and Siles, J. (2015) 'Flipping the Switch: Ensuring the Energy Sector Is Sustainable and Gender-Responsive', in Aguilar, L., Granat, M., and Owren, C. (eds.) *Roots for the Future: The Landscape and Way Forward on Gender and Climate Change*. Washington, DC: IUCN & GGC, pp. 203–288.

Swirski, B. (2002) *What is a Gender Audit*. Israel: Adva Center. Tel Aviv-Yafo.

UNESCO (2015) *UNESCO Science Report: Towards 2030*. Paris: UNESCO.

Wagner, H.M. (2011) 'The Bottom Line: Corporate Performance and Women's Representation on Boards (2004–2008)', *Catalyst*, 1 March [online]. Available at: www.catalyst.org/research/the-bottom-line-corporate-performance-and-womens-representation-on-boards-2004-2008/ (Accessed: 1 October 2020).

WISE (2017) *Women Employment in Urban Public Sector* [online]. Available at: www.wiseproject.net/downl/final_wise_project_report.pdf (Accessed: 1 October 2020).

5 Producing gender statistics at local level

The case of Mito-city, Japan

Keiko Osaki-Tomita, Reiko Gotoh, Miya Ishitsuka, and Yoshitaka Hojyo

Introduction

Gender inequality persists worldwide, depriving women and girls of basic rights and opportunities. To fight against the inequality, the United Nations (UN) has long played a pivotal role in raising awareness about gender issues and advocating for gender-sensitive policies. Most recently, the Sustainable Development Goals (SDGs) established by the UN in 2015 sets achieving gender equality and the empowerment of all women and girls as a stand-alone Goal 5 out of 17 goals. It also underscores that investments in gender equality can be a powerful spur to achieving all of the SDGs.

Gender inequality is often more pronounced in rural localities, especially in the country like Japan where its long history has shaped social norms and roles played by women and men. In order to advance gender equality for the nation, it is therefore imperative to understand gender relations at local levels, with an eye toward formulating effective policy interventions tailored for local features.

Evidence-based policy-making, based on available statistical evidence, is an approach which has received growing attention in public policy, even in the fields aiming at advancing the status of women and gender equality. The efficient use of statistics can validate the impacts of existing policy and promote the appropriate policy suited for a nation and localities. Hence, the UN set forth the Minimum Set of Gender Indicators to be used across countries and regions, for the national production and international compilation of gender statistics (United Nations, 2011). The Minimum Set is now fully aligned with the SDGs,

Existing research on gender statistics primarily concerns the scarcity or shortfalls of information to formulate effective evidence-based policies and programs at the national level. Little is known or investigated about the availability of gender statistics at a local level, and how the production and the use of gender statistics are different from those at the national government.

This chapter examines the availability of gender statistics at a local municipality office, using the example of the Mito-city government, Japan. The chapter begins with the review of gender policy and gender statistics in Japan. Second, it introduces recent attempts by the city government to formulate evidence-based policy-making in the areas of achieving gender equality, and elucidates the systemic challenges that prevent the production and use of gender statistics at the city.

DOI: 10.4324/9780367743918-6

The chapter also presents some unique findings from a citizen survey conducted by the city in 2018. Finally, it proposes several concrete plans to advance the statistical system of the city government for better production and use of local statistics.

Gender policy and statistics in Japan

History of gender policy in Japan

Japan has been always known as a country with a large gender gap. According to the 2020 Global Gender Gap Report, Japan ranks 121st of 153 countries, the worst among advanced economies. The main factors leading to this low rank are poor scores on political empowerment and on economic participation and opportunity for women (World Economic Forum, 2020).

Over the past decades, the Japanese Government has implemented institutional changes to reduce gender disparities, heightening public awareness on gender issues inherent in the country. The first postwar wave of institutional reforms came in the mid-1980s and led to the enactment of the Equal Employment Opportunity Law in 1985. The second wave came in the late 1990s, following the adoption of the Beijing Platform for Action in 1995, at the Fourth UN World Conference on Women, when the world's governments declared to fully commit to advancing gender equality. In Japan, rapidly dropping fertility rates and prolonged economic recession transformed gender equality into a political issue for economic and social revitalization, rather than treating it only as a human rights issue (Osawa, 2000). In 1997, the Government amended the Equal Employment Opportunity Law of 1985 to introduce mandatory measures for gender equality in recruitment, hiring, assignments, training, and promotion; these enforcements were lacking in the law when it was adopted (Gelb, 2003). Furthermore, the Basic Plan for Gender Equality 2000 was drafted in December 1996 as the fifth such national plan by the Headquarters for the Promotion of Gender Equality, with the prime minister acting as president. In line with this plan, the Basic Act for Gender Equal Society was enacted in 1999.

History of gender statistics in Japan

The Basic Act mentioned earlier requires local governments to establish basic plans for gender equality and to promote policies for realizing a gender equal society in a comprehensive and systematic manner. Establishing such basic plans is mandatory for prefectures and optional for municipalities. This requirement has amplified the importance of gender statistics for measuring and improving the performance of gender equality policies. Shortly following the Basic Act for Gender Equal Society of 1999, significant progress was observed in gender statistics in Japan in the early 2000s (Amano, 2004; Itoh, 2003). First, the Complaint Handling and Monitoring Specialist Committee of the Council for Gender Equality of Japan adopted gender statistics as a matter of priority and requested both national government agencies and local governments to improve the availability of gender statistics. Second, the document indicating the new direction of statistical administration of the Statistics Bureau of Japan demanded that data be collected and disaggregated by sex. Third, in academia, the Japan Society of Economic Statistics reinforced its efforts to mainstream a gender perspective in the statistical activities of national and local governments (Sugihashi, 2019).

Prompted by these developments, the Basic Plan for Gender Equality of Japan came to emphasize the need to promote gender statistics as a basic requirement for realizing a gender equal society. For example, the latest Fourth Basic Plan for Gender Equality of 2015 claimed that the government would work to gather, process, and provide gender-specific data, which are disaggregated by age and by prefecture as much as possible in various official statistical surveys and administrative statistics.

Limitations of gender statistics and gender studies in Japan

How much progress has been made in formulating local plans for gender equality and enriching local gender statistics? Although it was optional for municipal governments to draw plans for promoting gender equality, remarkable improvements have been achieved. At the end of October 2002, only 2.4% of the 3,240 municipalities in Japan had plans for promoting gender equality (Osawa, 2005). By 2018, the proportion jumped to 73.1% of 1,741 municipalities.[1]

As for gender statistics, reasonable amounts of data have become obtainable at a national level. Of the 52 quantitative indicators that the United Nations Minimum Set of Gender comprises, 38 indicators are available for Japan as of September 2020; that is, slightly fewer than France and Germany, but slightly more than Canada, Sweden, and the United States.[2] However, gender statistics for lower geographical levels remain far from sufficient. This is mainly because official statistics depend heavily on sample surveys, not on administrative records. Eventually, released official statistics based on sample surveys are classified only for eight main regions or 47 prefectures, or at best, large cities. This is because the margin of error becomes too large if data are disaggregated by municipality (Itoh, 2012). Administrative records covering the entire population are more suitable for disaggregation by municipality, but the statistical use of those records has been limited to date. Even the information on tax is not disclosed to the Statistics Bureau of Japan (Matsuda, 2012).

Moreover, anonymized data that contain municipality information are not readily available, though the amendment of the Statistical Act in 2007 made it possible to use the anonymized data of official surveys for general use. The number of gender studies in Japan has increased especially since the "Womenomics" policy was adopted by the former prime minister Shinzo Abe (e.g., Hara, 2018; Morikawa, 2016; Onozuka, 2016; Takeda, 2018). However, most studies rely on data that lack local gender-specific information and do not consider the gender gap at a local level. As a result, municipal governments have no choice but to obtain their own data or to gather evidence by themselves in order to plan and implement evidence-based gender-sensitive policies tailored for their local needs.

Gender policy and statistics in Mito-city

Mito-city is the capital of Ibaraki Prefecture, Japan, and located about 100 km northeast of Tokyo. The city, with the population of some 270,000, is primarily a regional commercial center and rich in history, nature, and culture. It witnessed the history of the Tokugawa Dynasty, resulting in many historical sites in the city. Mito also boasts of Kairakuen Park, one of the top three urban gardens in Japan.

Gender policy of Mito-city

Achieving gender equality in society has long been a policy priority of the city. Mito adopted the Declaration on Gender Equality at the City Council in 1996 and ranked third city in Japan that openly declared its commitment to advance the status of women. Subsequently, the city government issued the Basic Ordinance of Mito-city for Gender Equality in 2001 and stipulated to formulate a basic plan and implement it, together with citizens and business operators, in order to realize a gender-equal society. Since then, it has developed three Basic Plans of Mito-city for the Promotion of Gender Equality. The first Plan formulated in 2004 included the actions and programs involving all departments of the city government. Witnessing rapid societal changes, the second Plan developed in 2015 prioritized the actions to be implemented in the next five years. These Basic Plans of Mito-city were created based on opinion polls of citizens, together with being built on the experiences of implementing past national policies and programs of similar themes.

It was the city's Action Plan for the Promotion of Gender Equality in Workforce of 2017 that, for the first time, Mito-city attempted to introduce the evidence-based policy-making (EBPM) approach. The Plan had, among others, particular emphasis on promoting women's active participation in the workforce, and was a response to the National Act on the Promotion of Women's Participation and Advancement in the Workforce, stipulated in 2015. The time available for city officials to draft the Action Plan of Mito-city was too short to conduct a citizen survey, which would have captured the experiences of women in the city. Thus, they searched in vain for relevant gender statistics within the city government. Eventually, the dearth of data at the city level forced them to use the statistics owned by the national government as supplemental data in the preparation for the city's Action Plan. This made city officials realize the urgent need to enhance the statistical evidence base at the local level from a gendered perspective.

Statistical capacity of the city government

The quest for gender statistics in the city government revealed that, during the fiscal year 2018, the city's 57 public plans and programs were drawn in consultation with 20 surveys from which statistical data were accessible and available. However, inherent systemic issues were preventing the production and use of gender statistics at the city government. First, only nine out of 20 surveys conducted by the city collected the information on the sex of respondents. Hence, the gender disaggregation of data was not possible for the information gathered through those surveys. Second, the results of the surveys were used solely for planning purposes within the department concerned, and hardly shared with other offices of the city; that is, raw data were not made available to other offices for analyses. This can be seen as a waste of public goods and calls for the re-establishment of systems to facilitate fuller use of official data collected from citizens.

The lack of gender statistics at Mito-city is also evident when compared with the national government. Figure 5.1 shows the availability of 52 quantitative indicators set forth by the United Nations as Minimum Set of Gender Statistics, for Japan as a whole and Mito-city specifically. They are organized into five domains: economic structures and access to resources, education, health and related services, public life and decision-making, and human rights of women and child.

Overall, the Japanese central government could produce the data of 35 gender indicators as of April 2018, whereas Mito-city produced only 18, or about a half of that of the national government. Both national and city governments possess reasonable amount gender indicators relevant to economic dimensions (11 and 9, respectively). In contrast, the city is seriously lacking benchmark data to assess the situation of women and men in the areas of education and health. These huge gaps in the availability of data may be due to the absence of a mechanism that enables the central government to classify the data by smaller geographical regions, and disseminate them to municipal offices. Ultimately, the lack of statistical evidence prevents Mito-city from applying evidence-based policy-making in formulating gender-sensitive education or health programming.

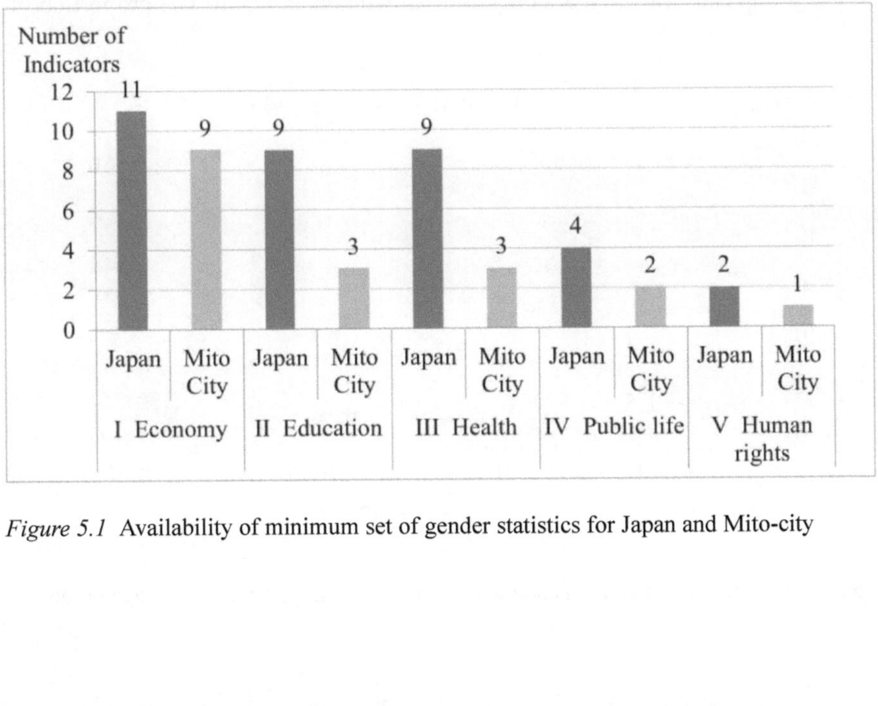

Figure 5.1 Availability of minimum set of gender statistics for Japan and Mito-city

Results from the 2018 citizen survey

In 2018, the city government conducted a citizen survey to prepare for the formulation of the Third Basic Plan of Mito-city for the Promotion of Gender Equality for adoption in 2020. The survey intended to explore, among other issues, the work patterns of and labor force participation of women and men, annual income earned, and how citizens strike a balance between work and life. The findings from this 2018 citizen survey demonstrate some features of gender relations unique to people of Mito-city.

Figure 5.2 suggests that, in Mito-city, there are striking differences between women and men in career development. The proportion of women and men who hold managerial positions increases as they age. However, the progression to managerial positions is much slower among women than among men. In their 50s, about one in two men holds a managerial position, whereas less than 10% of women are in that position. It seems that the gender gap in career development already starts widening when employees are in their 30s, possibly when many of them build a family. Only 2% of women and 12% of men are in managerial position in their 30s. The city plans to look closely into factors behind these results, as well as explore possible policy measures to narrow the gap.

Figure 5.3 presents the proportion of women and men who favor women continue to work even after they bear a child. On average, 62% of women and 55% of men consider that it is better for women to continue working, hence women showing somewhat stronger preference of making work and childrearing compatible. The highest figures were found among women aged 25–34, and among men aged 45–54, indicating 71% and 65%, respectively. These proportions decline gradually for both women and men as they become older.

Unexpected results found from the analysis are the lowest figures among youngest age group (i.e., 18–24 years old) for both women and men. The proportion of

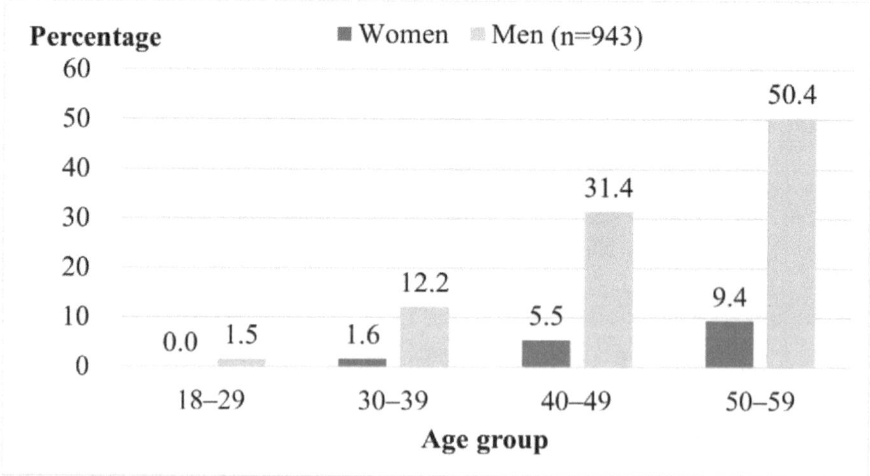

Figure 5.2 Proportion of women and men who hold managerial position by age group

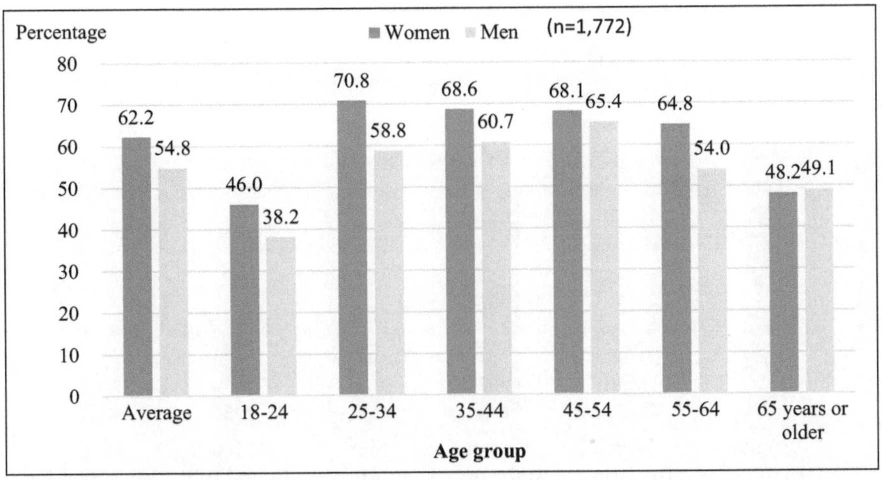

Figure 5.3 Proportion of women who consider that it is better for women to continue working after having a child

people who support the idea of women making a compatible life between work and childrearing accounted for only 46% among women aged 18–24, and 38% of men in the same age group. Prompted by these surprising findings, the city plans to investigate what made young people in the city have such conservative views with regard to women's life style.

Figure 5.4 shows an average time per day that married women and men engaged in unpaid domestic and care work. The data show that women in Mito-city spend 6.8 hours in domestic or care work on average, whereas the corresponding figure for men was only 2.2 hours. Hence, women spend more than three times longer than men undertaking unpaid domestic and care work. When samples are limited to full-time employees, the time spent for domestic and care work decreases by more than one hour for women, but it is still nearly three times longer when compared to men.

Another interesting finding is that there is no significant difference in the time spent for domestic and care work between liberal and non-liberal men. Here, being liberal means those people who consider that it is better for women to continue working even after having a child. Liberal men in the city spend two hours in domestic or care work per day, whereas non-liberal men are engaged in this work for 1.6 hours. It can be interpreted that even liberal-minded men spend as little time as non-liberal men engaged in domestic work. This result suggests that changing men's perspectives on gender relations is not enough to facilitate their participation in domestic or care work. There may be external factors such as long working hours or commuting time that may prevent men from sharing domestic work with women.

Thus, the aforementioned analyses reveal some features of women and men unique to the residents in Mito-city, demonstrating the usefulness of having

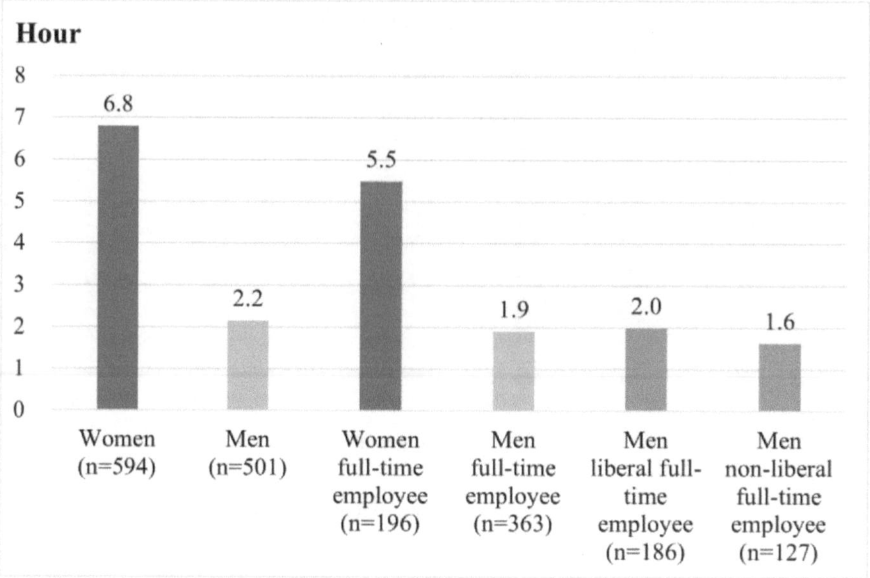

Figure 5.4 Average time spent in unpaid domestic and care work by married women and men

gender statistics and data at a local level. It should be noted, however, that one cannot expect such surveys to be conducted regularly or frequently. Most municipal governments not only have constrained budgets for research but also have limited technical knowledge to carry out citizen surveys.

The way forward

The quest for gender statistics and gender-based analyses using citizen surveys led the city government to realize the urgent need to further improve official statistics from a gendered perspective. Below are a range of concrete actions and strategies that the city proposes to improve gender statistics both in quantity and in quality.

First, the city suggests to scale up its effort to reach a wider audience, and improve communication between data producers and users. It can organize seminars and workshops targeting college students or young people, so as to sensitize them about gender relations and associated issues. Through such occasions, participants should learn that achieving gender equality is one of pressing local as well as global issues as committed in the SDGs. Using data and statistics as communication tools, the opportunities to discuss evidence-based policy-making should be also offered to younger generations who will define the future. Already, classes of this kind have been conducted at local universities in Mito.

Second, the city considers to widely promote open data principles and establish a new data hub which will allow easy access to existing official statistics by citizens and researchers, with a view to facilitating fuller utilization of statistics at the local level. In doing so, due attention should be paid to confidentiality so that the personal information of citizens not be revealed.

Indeed, an increasing number of municipal governments in Japan have implemented an open data principle by setting up, for example, an open data catalog site on its website. Mito-city government also opened the site. However, access to the site has been limited. One should bear in mind that creating an open data catalog site alone is not sufficient enough. It is expected that the city would increase its efforts to facilitate the better use of the public data by widely publicizing its existence among citizens and researchers. Furthermore, in order to promote the use of data beyond the municipal government, there is a need for municipal offices to standardize the data format they use, so as to allow easy comparison of data between municipalities. The national government may wish to take a lead in such tasks.

Third, it is of critical importance to enhance the data literacy of city officials as well. The city wishes to appoint a so-called data concierge, whose role is to properly guide officials with respect to how data can be interpreted and used for policy-making. Most municipal governments in Japan face rapid population aging and decline, resulting in serious financial difficulties. There is a growing demand for local governments to be more accountable in using financial resources and to provide a transparent rational for choosing policies and programs. Statistical data are the most objective evidence to justify such collective decision-making.

Conclusions

This chapter reviewed the quest for gender statistics by the Mito-city government in the context of evidence-based policy-making. It revealed a critical dearth of data and statistics at the local level, as well as the lack of a mechanism to fully utilize existing statistical resources. The analyses demonstrate the critical importance of having data at smaller administrative levels in order to capture unique gender relations of citizens who live in the local community. A local government without data cannot draw its future. It is of paramount importance for local governments to have its own data strategies and evidence base for policy-making so that no local citizen – whether women or girl – would be left behind. Investments in gender statistics would effectively improve local gender policy, contributing to the achievement of the SDGs and the realization of sustainable, gender equal society.

Notes

1 The number of municipalities declined during the period because of municipal mergers. The figures are as of June 14, 2019, and quoted from *Numerical Targets and Updated Figures of the 4th Basic Plan for Gender Equality* [Online]. Available at: www.gender.go.jp/about_danjo/seika_shihyo/pdf/numerical_targets_2019.pdf. (Accessed: 15 September 2020).

2 Available at: https://genderstats.un.org/#/data-availability. (Accessed: 15 September 2020).

References

Amano, H. (2004). Studies on gender statistics (Gender tokei ni kansuru tyosa kenkyu) (in Japanese). *Journal of the National Women's Education Center of Japan*, 8, pp. 81–91.

Gelb, J. (2003). *Gender policies in Japan and the United States: Comparing women's movements: Rights and politics*. New York: Palgrave MacMillan.

Hara, H. (2018). The gender wage gap across the wage distribution in Japan: Within- and between- establishment effects. *Labour Economics*, 53, pp. 213–229.

Itoh, S. (2003). Trends in gender statistics and gender statistics studies (Gender tokei and gender tokei kenkyu no dokou) (in Japanese). *Academic Trends (Gakujutsu no Doukou)*, 8(4), pp. 28–31.

Itoh, Y. (2012). Creating a collection of local gender statistics: Current status and challenges (Tiho gender tokeisyu no sakusei: genjyo to kadai) (in Japanese). *Statistics (Tokei)*, 63(5), pp. 2–8.

Matsuda, Y. (2012). The future of statistical reform in Japan: The possible statistical system facing the changing world where the concept of nation state is fading (in Japanese). *Journal of the Japan Statistical Society*, 41(2), pp. 341–354.

Morikawa, M. (2016). What types of company have female and foreign directors? *Japan and the World Economy*, pp. 37–38, 1–7.

Onozuka, Y. (2016). The gender wage gap and sample selection in Japan. *Journal of Japanese and International Economics*, 39, pp. 53–72.

Osawa, M. (2000). Government approaches to gender equality in the mid-1990s. *Social Science Japan Journal*, 3(1), pp. 3–19.

Osawa, M. (2005). Japanese government approaches to gender equality since the mid-1990s. *Asian Perspective*, 29(1), pp. 157–173.

Sugihashi, Y. (2019). Development of activities on gender statistics of the United Nations and Japan (Kokuren to Nihon no Gender Tokei Katsudo no Tenkai) (in Japanese). *Labour Survey (Rodo Tyosa)*, 587, pp. 4–9.

Takeda, H. (2018). Between reproduction and production: Womenomics and the Japanese government's approach to women and gender policies. *Gender Studies (Gender Kenkyu)*, 21, pp. 49–70.

United Nations (2011). *Report of the forty-second session, statistical commission*, Decision 42/102. E/2011/24-E/CN.3/2011/37.

World Economic Forum (2020). *Gender gap report 2020* [Online]. Available at: www3.weforum.org/docs/WEF_GGGR_2020.pdf (Accessed: September 2020).

Part 2

Target 5.4

Value unpaid care and promote shared domestic responsibilities

This second section focuses on *Goal 5, Target 5.4*, specifically *Indicator 5.4.1: Proportion of time spent on unpaid domestic and care work, by sex, age and location*. Highlighting the geographies of North America (Mexico, the USA, and Canada), and SSA (Ghana), the five chapters in this section interrogate the frequent presumption that women and girls are "made" for providing unpaid, informal care work, while best suited to undertake the many responsibilities of domestic work, whether in the home, community, or beyond. The interconnectivity of *Goal 5* with *Goal 2: Zero Hunger*, and *Goal 6: Clean Water and Sanitation* is emphasized.

DOI: 10.4324/9780367743918-7

6 Gender statistics, geospatial analysis, and Sustainable Development Goals

A case study of Mexico

Margarita Parás Fernández, Claudia Tello de la Torre, and Paulina Grobet Vallarta

Introduction

With 17 goals, 169 targets, and 232 indicators (of which 54 are gender-specific), the Sustainable Development Goals (SDGs) represent a historic global commitment to achieve gender equality by 2030. Gender equality and the empowerment of all women and girls is not only an explicit goal but also a driver for sustainable development in all its dimensions – from ending poverty and hunger, promoting prosperity and inclusive growth, and building peaceful, fair, and inclusive societies, to securing the protection of the planet and its natural resources (Women, U. N., 2018).

Geospatial technologies are regarded as an invaluable tool for addressing critical challenges related to the measurement and monitoring of the SDGs. Despite the growing recognition of the power of spatial data and analysis for monitoring the SDGs, applications specific to gender equality remain limited to date. There are few studies where Geographic Information Sciences (GIS) applications have been used to help reveal spatial patterns in gender equality, employment, and violence, as well as securing rights for women and girls (Leszczynski and Elwood 2015, 2011; Fluri, 2009; Nelson and Seager, 2008; Coluccia and Louse, 2004; Koskela, 1997). Geospatial information can yield new insights on these issues not ascertained from traditional data sources while providing important opportunities for addressing gender data gaps across the SDGs.

The Global Centre of Excellence on Gender Statistics (CEGS), a partnership between National Institute of Statistics and Geography (INEGI) and UN Women, was launched in September 2018 as a platform for collaboration, knowledge sharing, and innovation on gender statistics. The goal was to contribute to the implementation and monitoring of the 2030 Agenda and the SDGs. In its capacity as an Innovation Lab, the CEGS aims to play a key role in unlocking the potential use of geospatial information in order to yield new insights on gender statistics and contribute to building an interdisciplinary network of experts in this field.

The objective of this chapter is to share our initial experiences regarding our work on cross-cutting gender approaches, place-based and geospatial analysis. A comprehensive view of significant contributions to gender research is presented, focusing on the SDGs and policies. Our efforts are geared toward communicating and informing stakeholders and directing future research and policy.

DOI: 10.4324/9780367743918-8

The chapter is divided into three sections. The first section introduces the importance of the place or territory, and its relationship with social inequalities from a gendered perspective. Using geo-spatial frameworks, we identify the handling of these issues as part of an international agenda, addressing the objectives of sustainable development. The second section illustrates a research approach that aims to identify relevant dimensions, integrating policy analysis in the conceptual model and accompanying methodologies. The third section provides an empirical exercise, using the case of Mexico, to integrate conceptual frameworks and gender statistics focusing on indicators of women's labor and income, and their correlation with care-related activities in different municipalities. A geospatial WEB platform, designed for cartographic and geo-statistical representation and visualization, serves as an intermediary in the communication of specialized data using a territorial and gendered approach, for both stakeholders and users.[1] Final remarks summarize the main messages put forward in this chapter, pointing to the efforts of CEGS in advancing gender geo-statistics and research so no woman or girl is left behind.

Gender and a territorial/place-based approach to SDGs

The territorial approach combines a strategic orientation involving the concept of place. The development of regional policy has brought to light the fundamental premise of the territorial approach, through the planning, design, and implementation of policies based on place at different scales (Centre for Industrial Studies, 2015; O'Brien, Sykes and Shaw, 2015; Barca, McCann and Rodríguez-Pose, 2012). To adopt a territorial/place-based approach involves incorporating intangible capital based on a combination of specific local factors. Place-based policies interact with the socio-economic behaviors of a place, reflecting the relevance of context, the player's interactions, and the geographic spaces in which the processes take place (Barca, 2009).

Bradford (2004) illustrates that the place-based approach is the best solution for local problems and community actions. A place-based approach contributes to the adaptation of institutions through: considering local requirements or needs; fostering opportunities and emerging initiatives; looking for flexible options to choose from; and implementing policies that can be adjusted at the local, regional, national, or international level (Reimer and Markey, 2008).

This chapter assumes that the territorial and place-based approaches are interrelated, although they maintain their conceptual and operational independence. The UN Women's global monitoring report *Turning promises into action: Gender equality in the 2030 Agenda for Sustainable Development* (Women, 2018) states that:

> Gender equality and the empowerment of all women and girls is not only an explicit goal under the 2030 Agenda, but also a driver of sustainable development in all its dimensions: to end poverty and hunger, to promote fair and

> inclusive societies and to secure the protection of the planet and its natural resources.
>
> (p. 72)

Furthermore, the report focuses on the urgent need for better quantitative and qualitative measurements and indicators related to women's lives, activities, and safety at the national and local scales.

Gender inequality and the undervaluing of women's activities and priorities have been replicated in statistical records (United Nations Data Revolution, 2014). Transforming statistical records to capture gender inequality will "require a significant increase in the data available to individuals, governments, civil society, companies and international organizations to plan, monitor and be held accountable for their actions" (p. 4). The Thematic Research Network on Data and Statistics (TReNDS) (2020) reaffirms the importance of creating inclusive societies, pointing out that "without reliable and timely population data linked to location, we cannot ensure that everyone is counted and no one will be left behind" (p. 7). This statement acknowledges that, in order to develop the policies and programs needed to fulfill the 17 SDGs, traditional data sources need to address geographic and temporal challenges.

For example, considering the impact of the COVID-19 pandemic at the global and local scale, the international community is re-evaluating the goals established by the Agenda 2030. In particular, the United Nations Development Programme (UNDP) is working on a forward-looking response to this crisis, which is affecting all human development dimensions, offering a collaborative scheme in the report *Beyond Recovery: Towards 2030* (UNDP, 2020).

To meet these challenges, several initiatives aimed at strengthening the sustainability criteria and means of implementation through a range of policy directions were proposed in the *COVID-19 Global Gender Response Tracker* (UNDP and UN Women, 2020), including finding new ways of social inclusion and determining equal opportunities for women through addressing forms of employment, jobs, and use of time, and introducing social and technological innovations which support sustainability, specifically in education, health, and care systems.

To address these tasks, CEGS proposes three main lines of research. First, inter- and transdisciplinary modeling and geospatial tools for spotlighting gender and SDGs. Second, geospatial information and technology management, to provide insight into territorial/place-based gender inequalities as well as exploration of the potential of policy instruments to overcome them (digitalization and wider access to internet and ICT infrastructure). Third, the narrative and communication to make visible the gender dimension in the territory concerned, recognizing the importance of the temporal and spatial dimensions for the analysis of gender issues, such as women's economic empowerment and care systems, and the increasing violence upon women and girls.

Transdisciplinary research deals essentially with the development of conceptual and methodological frameworks that provide the ability to integrate a wide range of disciplinary perspectives to best understand gender equity and the many other socio-economic and environmental problems of our time. Transdisciplinary collaboration is essential in the study and practice of emerging complex socio-ecological systems (Holland and Sigmund, 1995). Building a consensus over common conceptual principles is attained through transdisciplinary collaboration, sustained by the integration of specialized disciplines (Parás and López-Caloca, 2017). Territorial/place-based and gendered approaches deal with social complexities, converging in the commitment to expand social players' agency and women's empowerment and rights for inclusive and sustainable development (Mujeres, 2016; Ranaboldo, Cliche and Serrano, 2015).

An important outcome of this research is to provide meaningful feedback for gender policy and action, establishing bridges among the explicit knowledge models of key stakeholders and the institutions in charge of generating statistical and geographic information, as well as those involved in decision-making.

Women and men in society: territorial dimensions of everyday life

Territorial/place-based and gendered approaches, as conceptual and operational frameworks, are designed to respond to local contexts and challenges in order to achieve equitable and inclusive growth in accordance with Agenda 2030. In this context, "the territory" is understood as the product of an emergent process of social construction; it is the result of a complex system of interactions, where social agents relate, learn from experience, and adapt accordingly. It integrates place-based knowledge and information relevant for planning and policy decisions (Parás and López-Caloca, 2017).[2]

Social players transform their cultural landscapes as they move through their daily lives. Temporal and spatial patterns have provided numerous lines of inquiry for geographers and interdisciplinary researchers, with topics ranging from racial, ethnic and gender segregation, formal and informal labor markets, to health and community cohesion. The territory has become a paradigm for understanding urban and regional processes that occur within them (Camagni, 2017, 2007a, 2007b, 2001; European Commission, 2005). To illustrate this, in 2001, the Organization for Economic Cooperation and Development (OECD) proposed that each region should have a specific territorial capital to provide a context for regional policy. The territorial dimensions, alongside the economic and social, represent the objectives that have been assumed in the EU sectoral policies prescribed in the "Territorial Agenda 2020" (European Commission, 2011). The United Nations Economic Commission for Latin America and the Caribbean (ECLAC, 2019) has proposed a territorial and gendered approach to development planning, recognizing that the concentration of population and production generates inequalities and disparities among and within countries, affecting rural and urban populations differently.

Inequalities are multidimensional and related to the capacity, access to income, assets, land, natural resources, infrastructure, and political and social structures characteristic of precise locations and territories. This is the case for women working in productive processes. Their access to differentiated markets is affected by economic and environmental dynamics, where cultural heritage and biodiversity influence the role and identities of women and girls (Mujeres, 2016).

An intersectional approach examines the categories of difference at various levels of analysis and looks for the interactions among them (Choo and Ferree, 2010). Inter- and transdisciplinary knowledge, and different sources – such as statistics, geo-spatial models, and existing data – are key to sorting out the intersectionality of inequalities that emerge from the gendered issues in the territory concerned. Gender mainstreaming has been implemented in development programs and is anchored in the SDGs; further, it has developed toolkits for capacity building.

Gender statistics has been recognized as a central tool for giving visibility to the different manifestations of gender inequalities, generating a disaggregated base of information by sex, while conducting surveys on time and of violence against women (CEGS, 2018a).

Gender-disaggregated data are needed to make women visible and to make informed policy. In this respect, gender is just one of many blind spots in many of the data that are produced; numbers on disability, or ethnicity for example, are also often lacking although they are essential for diagnosing and solving the inequalities and exclusion that drive poverty:

> Collecting data that more accurately reflects the experiences of women, and that provides policy makers with the raw materials for better, more gender sensitive policy, is one part of the gender data revolution. But there is a second objective: putting data in the hands of women themselves.
>
> (Badiee and Melamed, 2014; www.undatarevolution.org/2014/12/15/gender-data-revolution/)

Geo-spatial framework for advancing SDGs

The CEGS presents an integrative gender and geostatistical framework, at conceptual and operational levels, that responds to and monitors SDGs[3] at national, subnational, and local levels. The methodological approach is based on an inter- and transdisciplinary approach for territorial and place-based gender research and collaboration, supported by geospatial analysis of social processes, socio-economic infrastructure, and care and prevention services. Geo-spatial information and analysis enable the development of indicators at a local level for the diagnosis and evaluation of public policies and programs concerned with the economic empowerment of women while contributing to monitoring the advancement of the 2030 Agenda.

There are several studies that refer to geospatial analysis of social phenomena. The use of new technological tools facilitates the calculation of indicators, and the method and analysis of data collection. For example, given the recent advancements in Earth observations, it is now possible to get more frequent and more granular georeferenced data for: population estimates, human settlements, and infrastructure worldwide using gridded population datasets (POPGRID Data Collaborative, 2018). However, many geospatial and statistical studies have ignored place and gendered dimensions of such phenomena. Kwan (2013) has argued for the need to expand our analytical focus from the static residential spaces captured by census-based studies, to other places and times in people's daily lives.

To meet the CEGSs' purpose, a network of specialists in gender issues and Geographic Information Sciences (GIS) was set up as a collaborative team, supported by stakeholders' participation in the building of scenarios. This Knowledge Network (KN) (Parás and López-Caloca, 2017) provides the connection and interaction of specialists in gender issues and territorial and spatial analysis with those in charge of formulating and implementing public policy.

GIS technology and tools were adapted to manage data and information, expanding our knowledge through modeling. The main objectives of the geospatial framework are:

I To develop strategies and support questions which need urgent answers, based on qualitative and quantitative data and geo-statistics visualization
II To define sources of information and types of data required, including location, time, and scale features
III To use spatial analysis techniques and methods
IV To integrate an innovative technological platform using resources from open and developers' tools that enable users to perform analysis and make informed decisions

The methodology focuses on providing the user with a functional prototype for the analysis and visualization of multiple dimensions of gender inequities and women's empowerment. These are useful for the design and evaluation of public policy and programs. The CEGS is supported by an open web platform that links geospatial information, databases, and geospatial analysis tools via a system that enables the organization, integration, and communication of gender knowledge and information on women's empowerment and care (see Figure 6.1).

A conceptual model for integrating a gender perspective in public policies

Policy implications and strategies linked to the spatial and territorial analysis of gender issues

The effective design and implementation of gendered policies requires a commitment to build institutional capacity within the decision-making process (Kerselaers, Rogge, Vanempten, Lauwers and Huylenbroeck, 2013; Primdahl, Kristensen and Busck, 2013). Spatial analysis incorporates the complexity of the interaction and feedback that derives from both societies' and stakeholders' participation. This synthesis gives rise to a strategy aimed at incorporating a gendered approach to policy at multiple scales.

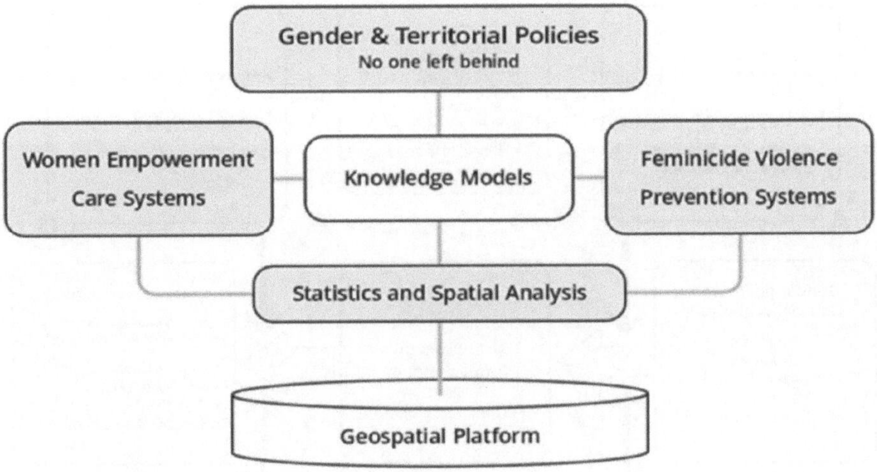

Figure 6.1 Knowledge base on gender statistics and geospatial analysis

The conceptual model developed (Figure 6.2) has implications for both research and political agendas since it acknowledges the factors that derive from institutional capacity (Görg, 2007). Situational analysis focuses on place-based gender inequalities, linking governance and administrative levels with policies that must be reoriented to incorporate gender and place dimensions. Figure 6.2 illustrates the interrelations between different dimensions of spatial analysis (socio-institutional and territorial) and reflects the design and implementation of gender policies at the national or regional level. It should be noted that solid lines represent the influence on the process and dotted lines represent the mutual change between dimensions (Zasada, Häfner, Schaller, van Zanten, Lefebvre, Malak-Rawlikowska and Zavalloni, 2017).

"A" describes the influence of different players and the institutional framework. "B" describes the role of networks and the creation of benefits. "C" describes the spatial and scale goal, with adjustments or mismatches in the effectiveness of policies. "D" describes territorial conditions and other assets that affect the generation of benefits by a policy. The role of the territorial context in "C" relates to the effects of the spatial policy goal mismatch between the objectives and the effects that influence the policy's efficiency and effectiveness. In "D," the correlation with the distribution of benefits is a function of the spatial variation of the actions. The aim of space policy is also the result of the design of the policy and the planning process in which the different actors are involved (between "A" and "C"). Regarding relationships "A" and "B," cooperation and collaboration improve decision-making. Anything that may exist between letters "C" and "D" introduces spatial targeting and vicinity effects that will influence spatial distribution when adopting a certain policy action. "B" and "D" reflect the variations between the processes in different places.

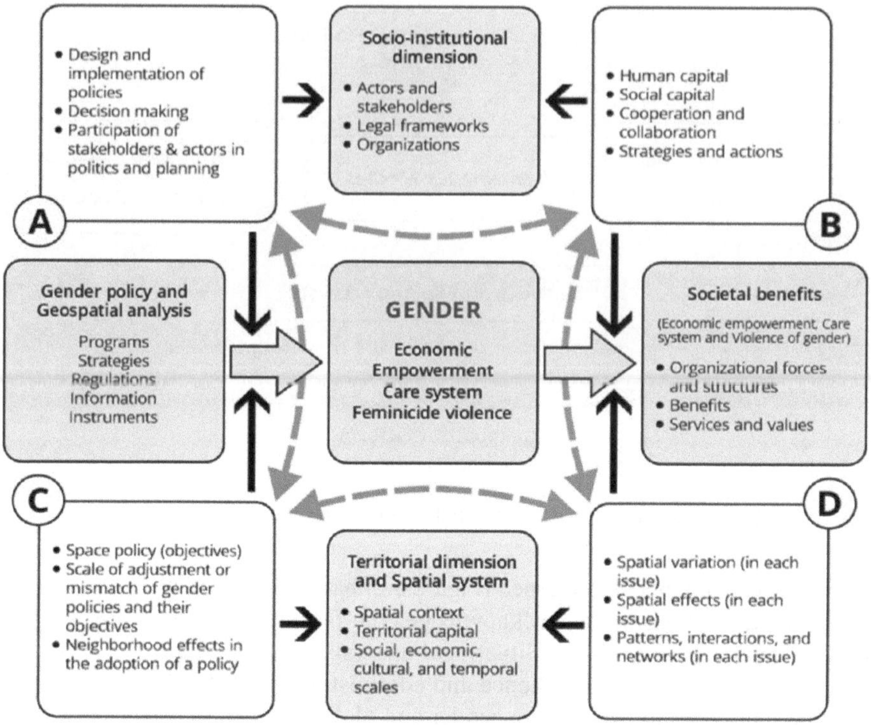

Figure 6.2 Conceptual model for Policy Integration

New networks emerge from these processes, and intermediate players facilitate the flow of necessary information and actions specific to the assignment of responsibilities, needs, knowledge requirements, and statistical information (Schomers, Matzdorf, Meyer and Sattler, 2015). Throughout this work, the importance of the participation of various players, decision makers, and specialists has been explored for the effective implementation of policies and programs. The responsiveness of geospatial and statistical analysis for gender research recognizes novel ways to quantify, spatialize, and communicate gender indicators for monitoring the compliance of the main objectives of Agenda 2030. The understanding of the analysis is enhanced by diverse forms of visualization (maps, graphs, tables, videos, diagrams, and qualitative data).

A systemic integration of gender policies

The integration of a method must be flexible in the face of multiple processes that occur within the same space (López-Caloca, 2011; del Carmen Reyes, López-Caloca, Lopez-Caloca, & Sánchez-Sandoval, 2014). Likewise, the variability of spatial processes generates imbalances in the scale and objectives of decision-making and in the spatial effects of policies.

Geospatial analysis can be developed from the specificity of places and generate relevant regions for the study of gender issues and indicators, such as those related to empowerment or violence against women. GIS analysis is useful to illustrate how territorial and socio-institutional aspects converge and enable the acknowledgment of patterns of information, spatial attributes, and the role of networks and actors in place. The spatial (local) peculiarities, and the mapping of the characteristics of the phenomena under study, make the system more complex, thus counteracting the effectiveness of policies or programs (Muller, Jolly and Vargas, 2010; Del Corso, Kephaliacos and Plumecocq, 2015).

An example of geo-statistical and gender analysis

This section puts to work the conceptual frameworks introduced above, illustrating the relevant elements for understanding the problematic issues in the territory of Mexico from a gendered perspective (CEGS, 2018–2020).

Economic empowerment of women and care systems

The relationship between economic empowerment and care systems has an impact on various aspects of women's lives, mainly in their ability to face their impoverished living conditions (Mateo and Rodríguez-Chamussy, 2016; Women, U. N., 2018; Orozco, 2018; Wong, 2012). Empowerment implies a process of change that enables women to choose between different alternatives that are strategic for their lives, whether through negotiation, exchange, and/or decision-making mechanisms. This process involves a transition from a situation where women have limited power, to another where their power is improved through: i) material, human, and social resources; ii) agency, including decision-making and negotiation; and iii) achievements, measured mainly through improvement in well-being.

There are three levels at which changes can facilitate the empowerment of women: personal, relational, and environmental (Kabeer, 1999). Personal changes refer to self-perception and trust, while that of relational implies changes in relationships with the family and the community. The environmental level refers to the changes taking place in the structure of social organizations and institutions.

On the other hand, care systems are understood as the "set of public and private actions that provide direct attention to the activities and basic necessities of daily life of people in a situation of dependency" (Uruguay, 2015: Art. 3). Care systems play a fundamental role in the distribution of people's use of time, especially in the case of women. Care policies include social security and protection systems (Mateo and Rodríguez-Chamussy, 2016). Beyond their formal organization in both tax and non-tax schemes, care systems are composed of subsystems, including non-formal schemes provided by households and their supportive social networks (Compton and Pollak, 2014; Dimova and Wolff, 2011), and paid household work for domiciliary services and private schemes.

Care systems permeate public health, education, social services, pensions, labor, transport, food, security, and urban planning policies (Durán, 2018), and are thereby associated with a range of different SDGs (Women, U. N., 2018). Societies with care systems that depend on the unpaid work of women in private homes via informal care networks contribute to the reproduction of unequal care loads affecting women and limiting their possibilities of empowerment (Elson, 2017).

The availability, quality, and accessibility of care services determine the conditions and provision of other basic social services. They influence the recognition, reduction, and redistribution of unpaid work for women. Care systems are part of the agenda of regional and international consensus, such as the Montevideo Consensus on Population and Development, the Copenhagen Consensus, the Declaration of Buenos Aires, and the SDGs. They are central to a transformation of public policies to achieve equality and development by 2030.

Gender and geo-statistical analysis: women's empowerment and relationship with care systems

The productive structure of a territory has different effects on the ability of women and men to generate income, as it influences the magnitude of labor participation between both sexes. The territorial approach is more effective in reducing development gaps in terms of gender. Promoting policies is not enough to increase the endowment of women's assets; it is necessary to consider their relative position to men, and the type of opportunities in their territorial context (RIMISP, 2015).

Economic empowerment includes access to one's own income, paid work, and social protection. Because of the unequal sex distribution of unpaid work and care systems, women have unequal access to these economic factors of empowerment, regardless of the economic strata (Women, U. N. and UNDP, 2017). The disproportionate care workload that women experience impacts their distribution of time, which is one of the root causes hindering the social and economic empowerment of women (Folbre, 1994; Kleven, Landais, Posch, Steinhauer and Zweimüller, 2019; Women, U. N., 2018). Care systems are key for reducing and redistributing women's unpaid work. Their importance is recognized in regional and international policy agendas. In particular, Agenda 2030 understands care systems as a strategic element in the transformation of public policies to achieve gender equality and boost development.

A geo-spatial approach recognizes the links between three central factors in the economic empowerment of women: labor supply (women´s capabilities, needs, and resources), labor demand (market), and care systems (social protection). Their interaction follows geo-spatial patterns. Visualizing these patterns is of great importance to identifying conditions that constrain women's empowerment, while informing action routes for successful public policies. The complexity of the analysis involves different data sources.

The accessibility of natural resources, infrastructure, and institutional norms and regulations enable the mobility and connectivity of people with other regions or territories. For women, access to basic social services, together with care and financial inclusion services, has a direct impact on their labor supply (see also Orozco, 2018; Orozco and Gammage, 2017). The case of Mexico illustrates the potential to explore the relevance and contribution of these factors.

Econometric model to predict women's labor participation

Women's labor force participation (WLFP) relates to the personal, household, and environmental factors that influence women's chances to enter the labor market. This is often presented in comparison to men´s labor force participation.

Taking into consideration the presence of small children as indicative of care needs in the household, it is possible to observe their influence in consumption, production, and use of time decisions in the household (App and Rees, 2009). Care needs to reduce women's availability of time to take part in the labor market. This pattern is replicated at the societal, macro, and local levels, and in municipal territories. The analysis of WLFP in 2,445 municipalities of Mexico shows that women's participation rate is associated with both their individual capacities – such as their education level – and the availability of care services for small children and persons with care needs in their locations.

To model the WLFP, aggregated data were used at the municipal level. Sources of information include the Inter-Census Survey (National Institute of Statistics and Geography (INEGI), 2015); the Economic Census (National Institute of Statistics and Geography (INEGI), 2015), and the Financial Inclusion Databases (National Institute of Statistics and Geography (INEGI), 2015) from the National Banking and Stock Commission (CNBV).

A Composed Index for Localities Accessibility, aggregated at the municipal level, was proposed as a spatial indicator (Frakes, Flowe and Sherrill, 2015).* It was constructed with the Travel Time Cost Surface Model (TTCSM), using data from the Continuum of Mexican Elevations 3.0, the National Roads Network 2017 (INEGI and IMT-SCT), and the set of data of Land use and Vegetation, scale 1:250 000 Series VI (INEGI). This variable showed a negative relation to economic participation. Greater isolation or lower accessibility impacts the proportion of women that participate in the labor force. The outcome could be associated with monetary and time costs correlated with women's labor participation, in cases where the economic activities are located far from their homes (Cogan, 1980).

Some of the main findings of the model include the rate of WLFP for the 15-year and older age group is 35% for the whole country, with an average income of $5,284 pesos monthly (National Institute of Statistics and Geography (INEGI), 2015). The average age is 39 years and the number of years of education is 9, which is equivalent to the completion of secondary school.

In female labor participation models, children aged 12 and under in the household have a negative effect on women's possibility to enter the labor market (Gammage and Orozco, 2008). The model at the municipal level confirms the pattern: a 10% increase in the proportion of children in this age group (from the total population) is associated with a 4% decrease in women's economic participation. It is important to highlight the fact that an increase in the provision of care services, as ten economic units per 100,000 inhabitants, is accompanied with an increment of 1.2% in WLFP.

Geospatial visualization of indicators related to Women's empowerment: a case study of Mexico

The prototype developed so far, and exemplified through the case study in Mexico, enables the visualization of main results obtained from the above model of WLFP, and the derived context for gender statistics compiled and aggregated at the municipal level.

As stated in Section 1 of this chapter, the objective of delivering a geospatial platform is to represent and communicate the spatial/temporal analysis of important variables for gender research. Once the models are explained, the user can build different scenarios by modifying the magnitude of the independent variables that can be informed by public policy, such as education level, fertility rate, healthcare service access, and accessibility of remote locations, while evaluating how these changes would impact women's empowerment (Figures 6.3 and 6.4).

A geospatial platform for gender research and analysis

A geospatial platform is proposed as a tool that allows users to navigate through an interactive digital menu to visualize and build scenarios for the study of variables related to gender and equity issues (Figure 6.3). Likewise, the technical design seeks to be a geospatial solution that expands the perspective of gender statistics on various topics. Its implementation involves the interaction of specialists with the purpose of providing geo-statistical information and modeling exercises from gender narratives and perspectives that enable understanding and dissemination of knowledge models, combining innovative techniques and development of space-time analysis.

The platform integrates a sample for the presentation, visualization, analysis, and construction of scenarios, as a first approach to its functionality. The use of statistical and econometric techniques allows the creation and/or modification of the selected variables, emphasizing the territorial perspective and connection with public policy schemes. The preliminary design of a visual interface integrates the results of the thematic analysis, in addition to providing geospatial information and contributing to the design and evaluation of public programs and policies. GIS is proposed for the development of an integral solution (Reyes, 2005). The basic premise of this methodology is outlined in Figure 6.5 and includes the conceptualization of a meta-system, system modeling, and technological solutions:

a) Meta-system: corresponds to the conceptualization stage in which the analysis of user requirements is considered, together with the specification of the

Figure 6.3 Women's economic empowerment and care systems

Source: +% of women that work (Scenario)

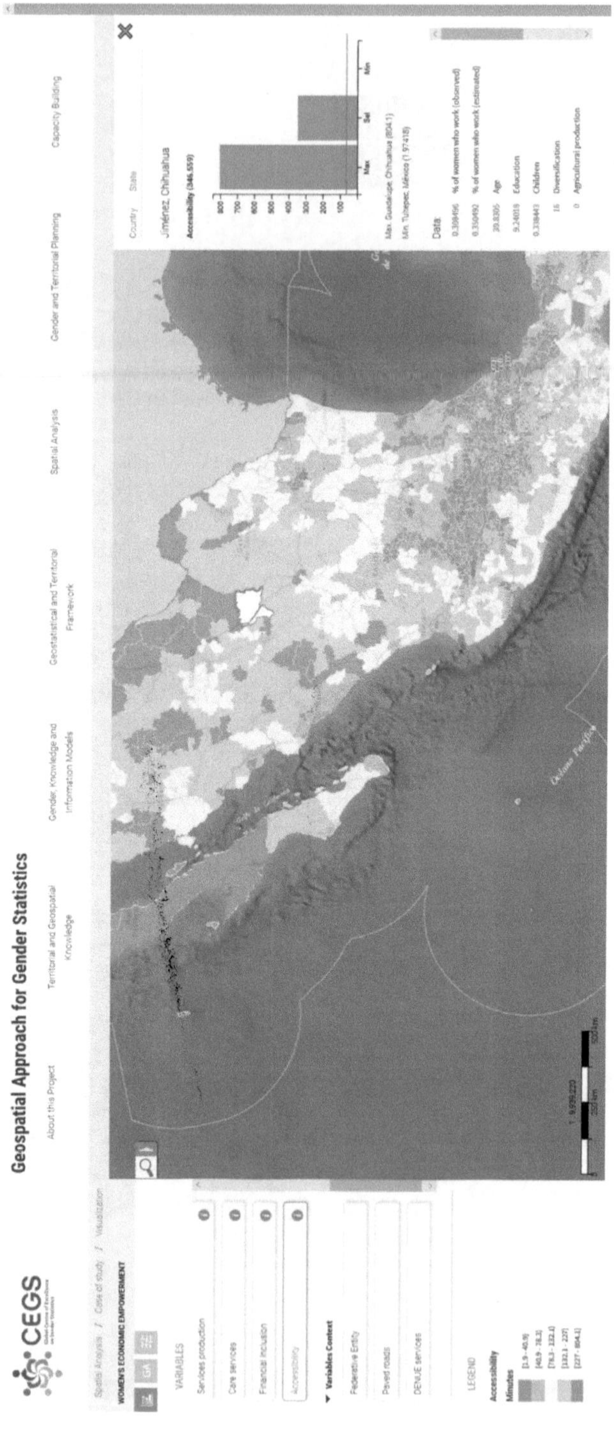

Figure 6.4 Women's economic empowerment and care systems

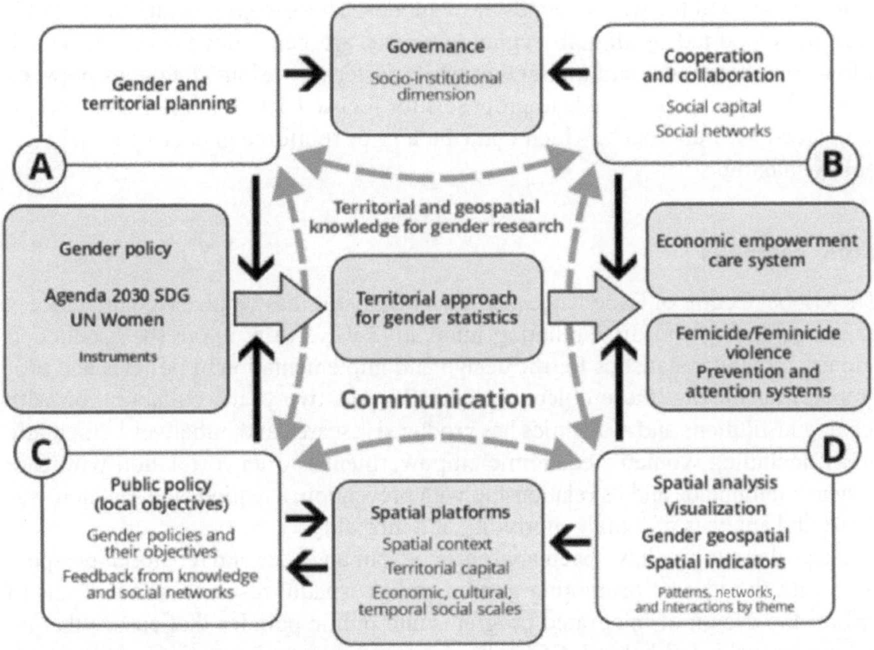

Figure 6.5 General platform framework

general knowledge frameworks for modeling, analysis and communication of geospatial content.

b) System modeling: corresponds to the selection of the models for the design of the solution, such as computational, geographic, visual, and cartographic models.
c) Technological solution: encompasses a technological strategy that meets the needs of the proposed solution, and includes the information and communication technologies available to integrate the technological platform. The first phase of a functional prototype was introduced with the CEGS, UN Women, and INEGI working group and presented among various users and major players.
d) The prototype has been developed with open source software. It is hosted within a WEB platform that allows the incorporation of spatial information systems, geospatial analysis tools, and web content in a dynamic interactive system, guiding the organization and dissemination of knowledge and information on women and gender issues.

The role that communication plays in the scheme depends not only on the flows but also on the structure, quantitative and qualitative content of the messages generated during the design and operational processes interchanged with end users. Therefore, the context of interaction influences the individuals' processing of meanings, and the way it relates these modes. These exercises and methodologies are supported by digital technologies that are being developed and which allow communication and interaction while fostering various dialogues between users. This approach can lead to proposals for focused action in the framework of the place-based approach, which contribute to or reinforce gender and territorial policy-making.

Final remarks

The Global Centre of Excellence on Gender Statistics has emphasized the process of strengthening and disseminating innovative ways to promote the production and use of gender statistics for the design and implementation of policies and programs. In achieving these objectives over the past two years, collaboration with leading institutions and academics has produced research and initiatives in thematic areas, including women's economic empowerment and its correlation with care systems; feminicide and its relationship with prevention, attention, and sanction systems; and analysis of female morbidity and mortality.

These three topics have been approached from a gender and territorial perspective, with the idea of promoting gender and geospatial research and analysis to inform the design of integrated programs and public policies that ensure that no woman or girl is left behind. CEGS has a special interest in continuing the development of new research in the context of the impact of COVID-19 on the lives of women and girls.

Many challenges persist in the monitoring of gender components of the SDGs making up the 2030 Agenda:

- The need for a more robust gender-responsive follow-up and accountability framework, designed to effectively measure and document inequalities between women and men, and reinforcing approaches to intersectional inequalities.
- The development of suitable gender indicators and statistics, in order to inform decisions, guide efforts of both governmental and non-governmental agencies, determine progress based on an adequate baseline, and monitor whether objectives are met within established timeframes.
- The need to increase gender indicators, specifically related to environmental issues, that integrate conceptual frameworks and methodologies in most of the SDGs, including global common standards, classifications, and interoperability systems.
- The monitoring of territorial/place-based and gender approaches, which are key to meeting the above challenges.

Little research has been done in the creation of international gender normative frameworks, or in designing conceptual approaches that are capable of operationalizing concepts for gender statistical purposes. Moreover, only a few studies integrate the use of sociodemographic and economic statistics with a gendered perspective, reinforced with geographical information characterized by a territorial/place-based methodology. The collaborative work undertaken by GCGS has assisted in building a spatial model not only capable of georeferencing gender indicators but also visualizing them through a technological geosystem. The transdisciplinary network has generated knowledge, information, and data analysis – all of which has responded to the requirements specific to users and policy makers, while communicating them in a visual platform.

CEGS is working in close collaboration with INEGI in order to migrate the geospatial prototype to a Geo Web platform. Here, it will influence the perception and understanding of political alternatives for decision-making, through sharing and promoting new research, building specific conceptual frameworks and methodologies, and using statistical and geographic information.

As mentioned by Santaella (2020) at the United Nations World Data Forum

> National Statistics Offices (NSO) and international organizations, have the challenge to provide information with different granularity to make gender gaps visible at different levels, (and) . . . the importance of providing the right communication to promote the use of this information by society and decision makers using visual displays and building context around information.

According to Hawkesworth (1988), the challenge for those working in the public sector is to interactively combine policies, social perspectives, and specialized knowledge on gender issues. That is, the sum of the technical know-how and the knowledge and experience at both the local and the political levels will contribute to problem-solving in a more inclusive form of participatory governance.

There are still some challenges ahead in terms of developing practical and accessible tools and harnessing the use of digital and open sources technologies.

CEGS continues to fulfill its mission to strengthen the discussion about gender inequalities and to close the gap in understanding the lives of women and girls through assuming the importance of targeting topics using disaggregated data to produce gender geo-statistics for monitoring 2030 Agenda SDGs. Collaboration among users and decision makers is crucial to sharing the lessons learned and prioritizing urgent and emergent topics, as well as identifying what is needed and what must be communicated to inform local policies and programs.

Finally, gender approaches and territorial and geospatial analysis are especially valuable for the CEGS when responding to the impact of crisis, such as the COVID-19 pandemic and climate change, on the lives of women and girls.

Notes

1 The WEB platform prototype has been developed with the collaboration of the Directorate General of Geography and Environment DGGMA-INEGI, Mexico.

2 Using a place-based approach is about bringing together citizens in a place to address the complex needs of communities by harnessing the vision, resources, and opportunities in each community.

3 The COVID-19 pandemic and the measures taken to overcome it will have a decisive impact on the general framework and opportunities for achieving the SDGs and climate targets. Sustainable development policy in crisis mode. Statement by SDSN Germany. June 2020.

* *This chapter arises as part of the discussion and research project "Gender and geospatial analysis" coordinated by CEGS. Special recognition is given to the transdisciplinary team of specialists: Margarita Parás; Mónica Orozco; Claudia Tello; Karla Ramírez; Juan Manuel Nuñez. Paulina Grobet oversaw the general coordination as CEGS Coordinator and Cynthia Rodriguez participated as technical assistance of the CEGS. The opinions expressed in this document are the responsibility of the authors.*

References

Apps, P., & Rees, R. (2009). *Public economics and the household.* Cambridge: Cambridge University Press.

Badiee, S., & Melamed, C. (2014). *Making the data revolution a gender data revolution* [Blog post]. Un Data Revolution. Retrieved December 15, 2014, from www.undatarevolution.org/2014/12/15/gender-data-revolution/

Barca, F. (2009). *Agenda for a reformed cohesion policy.* Brussels: European Communities.

Barca, F., McCann, P., & Rodríguez-Pose, A. (2012). The case for regional development intervention: Place-based versus place-neutral approaches. *Journal of Regional Science*, 52(1), 134–152.

Bradford, N. (2004, February). Place matters and multi-level governance: Perspectives on a new urban policy paradigm. Paper presented at the McGill Institute for the Study of Canada Annual Conference, Montreal, Quebec.

Camagni, R. (2001). *Policies for spatial development.* Paris: OECD.

Camagni, R. (2007a). Towards a concept of territorial capital. Joint Congress of the European Regional Science Association and ASRDLF, Paris.

Camagni, R. (2007b). Territorial development policies in the European model of society. POLIMI Publications Catalog. IRIS Polytechnic of Milan, Italy (pp. 129–144).

Camagni, R. (2017). Regional competitiveness: Towards a concept of territorial capital. *Seminal studies in regional and urban economics* (pp. 115–131). Cham: Springer.

Choo, H. Y., & Ferree, M. M. (2010). Practicing intersectionality in sociological research: A critical analysis of inclusions, interactions, and institutions in the study of inequalities. *Sociological Theory*, 28(2), 129–149.

Cogan, J. (1980). Teaching economic decision making in the intermediate grades. *Peabody Journal of Education*, 57(3), 154–159.

Coluccia, E., & Louse, G. (2004). Gender differences in spatial orientation: A review. *Journal of Environmental Psychology*, 24(3), 329–340.

Compton, J., & Pollak, R. A. (2014). Family proximity, childcare, and women's labor force attachment. *Journal of Urban Economics*, *79*, 72–90.

Centre for Industrial Studies. (2015). Territorial Agenda 2020 put into practice.

del Carmen Reyes, M., López-Caloca, A. A., López-Caloca, F., & Sánchez-Sandoval, R. (2014). Geocybernetics as a tool for the development of transdisciplinary frameworks. In Modern Cartography Series (Vol. 5, pp. 33-42). Academic Press.

Del Corso, J. P., Kephaliacos, C., & Plumecocq, G. (2015). Legitimizing farmers' new knowledge, learning and practices through communicative action: Application of an agro-environmental policy. *Ecological Economics*, 117, 86–96.

Dimova, R., & Wolff, F. C. (2011). Do downward private transfers enhance maternal labor supply? Evidence from around Europe. *Journal of Population Economics*, 24(3), 911–933.

Durán, M. Á. (2018). *La riqueza invisible del cuidado* (Vol. 30). Spain: Universitat de València.
Economic Commission for Latin America and the Caribbean. (2019). *Planning for sustainable territorial development in Latin America and the Caribbean*. Santiago de Chile: ECLAC.
Elson, D. (2017). A gender-equitable macroeconomic framework for Europe. *Economics and austerity in Europe: Gendered impacts and sustainable alternatives* (pp. 15–26). London: Routledge.
European Commission. (2005). *Territorial state and perspectives of European Union*. Paris: European Commission.
European Commission. (2011). *Territorial agenda 2020: Towards an inclusive, smart and sustainable Europe of diverse regions*. Retrieved from https://ec.europa.eu/regional_policy/sources/policy/what/territorial-cohesion/territorial_agenda_2020.pdf
Fluri, J. L. (2009). Geopolitics of gender and violence 'from below'. *Political Geography*, 28(4), 259–265.
Folbre, N. (1994). *Who pays for the kids?: Gender and the structures of constraint* (Vol. 4). New York: Taylor & Francis.
Frakes, B. T., Flowe, T., & Sherrill, K. R. (2015). National park service Travel Time Cost Surface Model (TTCSM). *Natural resource report NPS/NRSS/NRR: 2015/933*. Fort Collins, CO: National Park Service.
Gammage, S., & Orozco, M. (2008). *El trabajo productivo no remunerado dentro del hogar: Guatemala y México. Serie Estudios y Perspectivas*. Ciudad de México: Sede Subregional de la CEPAL.
Global Centre of Excellence on Gender Statistics (CEGS). (2018a). *Strategic plan, 2018–2020*. Mexico: UN Women.
Görg, C. (2007). Landscape governance: The "politics of scale" and the "natural" conditions of places. *Geoforum*, 38(5), 954–966.
Hawkesworth, M. E. (1988). *Theoretical issues in policy analysis*. New York: SUNY Press.
Holland, J. H., & Sigmund, K. (1995). Hidden order: How adaptation builds complexity. *Nature*, 378(6556), 453–453.
Kabeer, N. (1999). Resources, agency, achievements: Reflections on the measurement of women's empowerment. *Development and Change*, 30(3), 435–464.
Kerselaers, E., Rogge, E., Vanempten, E., Lauwers, L., & Van Huylenbroeck, G. (2013). Changing land use in the countryside: Stakeholders' perception of the ongoing rural planning processes in Flanders. *Land Use Policy*, 32, 197–206.
Kleven, H., Landais, C., Posch, J., Steinhauer, A., & Zweimüller, J. (2019, May). Child penalties across countries: Evidence and explanations. *AEA papers and proceedings* (Vol. 109, pp. 122–26). Cambridge, MA: National Bureau of Economic Research.
Koskela, H. (1997). "Bold walk and breakings": Women's spatial confidence versus fear of violence. *Gender, Place and Culture*, 4(3), 301–319.
Kwan, M. P. (2013). Beyond space (as we knew it): Toward temporally integrated geographies of segregation, health, and accessibility: Space-time integration in geography and GIScience. *Annals of the Association of American Geographers*, 103(5), 1078–1086.
Leszczynski, A., & Elwood, S. (2011). Privacy, reconsidered: New representations, data practices, and the geoweb. *Geoforum*, 42(1), 6–15.
Leszczynski, A., & Elwood, S. (2015). Feminist geographies of new spatial media. *The Canadian Geographer/Le Géographe canadien*, 59(1), 12–28.
López-Caloca, F. (2011). *Un aporte teórico: el prototipo geomático* (Doctoral dissertation, Tesis para obtener el grado de Dr. en Geomática). Centro de Investigación en Geografía y Geomática "Ing. Jorge L. Tamayo" AC, México.

Mateo Díaz, M., & Rodríguez-Chamussy, L. (2016). *Educación que rinde: mujeres, trabajo y cuidado infantil en América Latina y el Caribe*. Resumen.

Muller, P., Jolly, J. F., & Vargas, C. S. (2010). *Las políticas públicas*. Bogota, Colombia: Universidad Externado de Colombia.

Mujeres, O. N. U. (2016). El progreso de las mujeres en el mundo 2015–2016. Transformar las economías para realizar los derechos. Resumen. *Revista Estudios Feministas*, 24(2), 589–614.

National Institute of Statistics and Geography (INEGI). (2015) Intercensal Survey 2015. Retrieved from: http://en.www.inegi.org.mx/programas/intercensal/2015/

Nelson, L., & Seager, J. (Eds.). (2008). *A companion to feminist geography*. Cambridge, MA: John Wiley & Sons.

O'Brien, P., Sykes, O., & Shaw, D. (2015). The evolving context for territorial development policy and governance in Europe: From shifting paradigms to new policy approaches. *L'Information géographique*, 79(1), 72–97.

Orozco, M. (2018). *Agenda de investigación para la incidencia en políticas relacionadas con los cuidados y la oferta laboral de las mujeres*. Mexico City: México: Oxfam-GENDERS.

Orozco, M., & Gammage, S. (2017). *Cash transfer programmes, poverty reduction and women's economic empowerment: Experience from Mexico*. Geneva: International Labour Organization.

Parás, M., & López-Caloca, F. (2017). Transdisciplinar meta-design for geomatics applications. *Proceedings of the 21st: World multi-conference on systemics, cybernetics and informatics: WMSCI* (Vol. 1, pp. 48–53). Mexico City: National Council of Science and Technology.

POPGRID. (2018). *POPGRID data collaborative*. Retrieved from www.popgrid.org/

Primdahl, J., Kristensen, L. S., & Busck, A. G. (2013). The farmer and landscape management: Different roles, different policy approaches. *Geography Compass*, 7(4), 300–314.

Ranaboldo, C., Cliche, G., & Serrano, C. (2015). Enfoque territorial para el empoderamiento de las mujeres rurales. *América Latina y el Caribe*.

Reimer, B., & Markey, S. (2008). Place-based policy: A rural perspective. *Community Research Connexions*.

Reyes, M. D. C. (2005). Cybercartography from a modeling perspective. *Modern Cartography Series*, 4, 63-97.

RIMISP, C. L. P. E. D. R. (2015). *Agricultura familiar y circuitos cortos en Chile: situación actual, restricciones y potencialidades.* Santiago, Chile: Publicación de la Oficina de Estudios y Políticas Agrarias del Ministerio de Agricultura, Gobierno de Chile.

Santaella, J. (2020). The power of granular data: Integration of geospatial information and gender statistics. Session TA03.01. Mind the Gap: Assessing Progress towards Gender Equality with Innovation Approaches, UNWDF.

Schomers, S., Matzdorf, B., Meyer, C., & Sattler, C. (2015). How local intermediaries improve the effectiveness of public payment for ecosystem services programs: The role of networks and agri-environmental assistance. *Sustainability*, 7(10), 13856–13886.

Thematic Research Network on Data and Statistics (TReNDS). (2020). Leaving no one off the map: A guide for gridded population data for sustainable development. New York, NY: United Nations.

United Nations Data Revolution. (2014). *A world that counts: Mobilising the data revolution for sustainable development*. Geneva: United Nations, Independent Expert Advisory Group on a Data Revolution for Sustainable Development.

United Nations Development Programme. (2020). *Beyond recovery: Towards 2030*. Retrieved from www.undp.org/content/undp/en/home/librarypage/hiv-aids/beyond-recovery-towards-2030.html

United Nations Development Programme and UN Women. (2020). *COVID-19 global gender response tracker*. Retrieved from https://data.undp.org/gendertracker/

Uruguay. (2015). Law no. 19353, art. 3-B. *Normativa y avisos legales del Uruguay*. Retrieved from www.impo.com.uy/bases/leyes/19353-2015

Women, U. N. (2018). Why gender equality matters across all SDGs. *Turning promises into action: Gender equality in the 2030* (p. 337). UN Women. New York, NY: Uninted Nations.

Women, U. N., & United Nations Development Programme. (2017). *Investing in social care for gender equality and inclusive growth in Europe and Central Asia: Policy brief 2017/01*. New York: UN Women-UNDP.

Wong, Y. N. (2012, November). World development report 2012: Gender equality and development. *Forum for development studies* (Vol. 39, No. 3, pp. 435–444). Abingdon, UK: Routledge.

Zasada, I., Häfner, K., Schaller, L., van Zanten, B. T., Lefebvre, M., Malak-Rawlikowska, A., ...& Viaggi, D. (2017). A conceptual model to integrate the regional context in landscape policy, management and contribution to rural development: Literature review and European case study evidence. *Geoforum*, 82, 1–12.

7 Understanding women's unpaid work and domestic work

Using Photovoice to capture immigrant carer-employee experiences in southern Ontario, Canada

Zahra Akbari and Allison Williams

Introduction

Many countries around the world are still grappling with how to best measure SDG 5.4.1: Unpaid Care and Domestic Work (United Nations, 2020), often incorporating time-use questions into the census or labor force surveys, for example. Other countries have been augmenting such traditional population-based data with the collection of time-use data in the field, using structured interviewing techniques or time-diary approaches. For the most part, data measuring unpaid care and domestic work have primarily been examined temporally, along the axis of time only. As geographers, we argue that both time and space are important to measure given that many tasks incorporated into SDG 5.4.1 can only be realized across space. For example, one of the most common caring tasks in Canada is assisted transport, moving the care recipient from A to B to C – whether that be for healthcare appointments, accessing goods, or visiting (Dardas, 2019). Capturing such spatial and temporal data can be realized using a number of methodological approaches rarely used in the measurement of unpaid care and domestic work, such as global positioning systems (GPS) paired with time-use diaries, which can then be mapped using geographical information systems (Kwan, 2000). Another novel methodological approach rarely used in capturing spatial and temporal data specific to unpaid care and domestic work in the micro-spaces of home, where most care giving and domestic work takes place, is a range of qualitative approaches, such as walk-along interviews (Evans & Jones, 2011) and Photovoice (Ilagan et al., 2020).

Drawing on the social geographical literature concerned with the home environment, particularly the micro-spaces of home, this chapter uses Photovoice as a qualitative methodology that employs qualitative interview narratives based on photos captured by research participants. It provides a nuanced look at the experience of the spatial and temporal tensions inherent in the micro-spaces of the home for Iranian immigrant carer-employees in southern Ontario, Canada. Carer-employees (CEs) are family caregivers who provide unpaid care and support to their family, friends, and/or neighbors while working in the paid labor force at the same time (Ramesh, Ireson, & Williams, 2017). Feelings of isolation and depression are common among CEs working from home, especially those who already have a limited social network such as immigrant CEs (Spitzer et al., 2003).

DOI: 10.4324/9780367743918-9

Photovoice provides a unique methodological approach for understanding unpaid care and domestic work, documenting the unequal division of labor in the provision of unpaid care and domestic work. Results illustrate the need to establish and accelerate progress on the equal division of labor between the sexes when addressing unpaid care and domestic work (UN Women, 2015). Doing so will contribute to meeting women's well-being, given the impact that the equal division of care and domestic labor will have on their economic, social, health, and quality of life, among other outcomes.

The primary aim of this chapter is to understand the spatial-temporal tensions that Iranian immigrant carer-employees experience as they provide informal, unpaid care to an elderly family member while working from home in the context of urban southern Ontario, Canada. As with many immigrants to the west, caring for elderly family members remains a private, family responsibility, that is most often realized by the women in the family. Although all participants had lived in Canada for more than ten years at the time of the research, they were still very much living the traditional gendered roles that continue to be commonplace in Iranian society today. The carer-employee sample managed this caring responsibility for elderly family members while simultaneously being responsible for domestic work and conducting paid work from home. Given the vast amount of time spent at home for these carer-employees who provide family eldercare, engage in unpaid domestic work, and conduct paid work in the home, the micro-spaces of the home become central to their ability to manage their multiple roles. These micro-spaces present multiple spatial and temporal tensions in CEs' everyday life. The methods and results are now presented, the latter which are discussed within the framework of SDG 5.4.1.

Methods

Given the focus on examining the relationship between the temporal and spatial experience of unpaid care, domestic work, paid work, and home as a place, a qualitative research approach was employed (Hay, 2010). Research ethics was secured before recruitment began (MREB 2017–106). The study took place in three Canadian cities in the province of Ontario: Toronto, Hamilton, and Waterloo. Using purposive sampling, four (n=4) Iranian immigrant home owner female CEs who live in multi-generational households and conduct paid work from home were recruited for the research. All were middle-aged females taking care of an elderly family member in the same home where they simultaneously conducted their paid work. Given the Photovoice methodology, it was also important for CEs to be familiar with photography to some extent. In the end, one participant was enrolled from Toronto, two from Hamilton, and one from Waterloo. Following the Photovoice methodology, participants were asked to answer a specific set of questions using their self-captured photos (Wang & Burris, 1997). These photos were then used in combination with audio-taped face-to-face interviews (Wang & Burris, 1997). Photovoice is reputed for giving "voice" to the marginalized population of concern and, in this project, has allowed Iranian immigrant carer-employees to express themselves through talking about their experiences based on their self-captured photos.

The use of Photovoice was favored by the participants for several reasons, including the ability for participants to freely discuss their hardships. Although opening up to a stranger about one's personal struggles is a cultural taboo in Iranian culture, using the photographs as the center of discussion in the face-to-face interviews led to building rapport between the participants and the researcher. Accordingly, participants compared the research process to a therapy session. Another advantage of using the Photovoice methodology was its complex nature, where photos were used as rich data sources reflecting participants' emotions and experiences in the best way possible. CEs reported described feelings of honor and worth following participation in the research.

Photovoice was used over the course of two sessions. At the first meeting, the Photovoice methodology was explained to the participants, with a list of questions provided, together with instructions to take at least one photo for each question. Three of the questions were about participants' experiences in their caring and working environment, while the other three focused on their home environment and the physical assets they preferred to either change or keep. Participants were told that anything and any place could be the subject of their photos, including the micro-spaces and places of the home environment that they spent most of their time.

Following the first meeting, participants were given two weeks to take their photos. Participants presented their photos at the second meeting, where they discussed their photos with respect to the questions asked. Each of the photos was explained in detail during the second session, via one-on-one semi-structured audio-taped interviews. Photos taken by the participants were discussed in the interview, and specific photos were chosen that best reflected participants' experiences and emotions (Wang & Burris, 1997). Participants' photos were considered as the primary source of information, together with the narrative associated with each photo. All of the interviews were conducted in Persian, which assisted Iranian carer-employees in discussing their experiences as well as their photos.

Data analysis was initiated after the two-month data collection period. All of the transcripts were first translated into English (from Persian). The transcripts were then analyzed using thematic content analysis, where data were examined for patterns or "themes" within the data (Braun & Clarke, 2006). Using NVivo software, thematic analysis was performed. The photos taken by participants provide a deep understanding of the phenomenon of concern.

Results

Two main themes were identified using the Photovoice methodology: spatial and temporal tensions. Each of these themes consisted of several sub-themes (Table 7.1). Identified tensions greatly affected the physical, psychological, and social well-being of the immigrant carer-employees. It was learned, through the process of data analysis, that the most repeated word by Iranian carers in their interviews was "stress." The various challenges experienced by immigrant carer-employees were reflected in their selected photos, which are shown below alongside their quotes.

Table 7.1 Immigrant carer-employees major tensions.

Tensions	*Sub-categories*
Spatial	• Safety • Single or multi-level distribution • Size and layout • Furniture • Ambient features • Structural components • Heating ventilation, and air conditioning (HVAC) and electrical systems • Privacy
Temporal	• Monitoring and control• Time involved in daily activities

Spatial tensions

Spatial tensions were experienced by CEs due to having unsuitable spatial characteristics in the physical home environment. CEs struggled with these spatial tensions making their caring, working, and daily activities of living harder, longer, and more frustrating. As a result, their physical and psychological health was negatively affected. Within this theme, a number of sub-themes were found, including safety, housing type, size and layout, furniture, ambient features, privacy, and structural components. The following sections provide a description of each of these sub-themes.

Safety

Ensuring safety of the care recipient was as one of the main concerns and sources of tension for CEs, especially when they were busy with paid work. This was specifically the case for CEs who had care recipients diagnosed with Alzheimer's disease, and who were not fully aware of their surroundings. CEs experienced several negative consequences due to these tensions, such as not being able to concentrate on their paid work, not sleeping well and constantly feeling stressed. This brought serious negative physical and psychological health consequences for them, especially for those who were not self-employed and did not have a flexible working schedule. Accordingly, one of the participants expressed her constant concern for her mother's safety, even at night:

> I am constantly worried about her even in the midnight. You know, one time she went down from the stairs in that darkness and I was very worried, although we had changed the entrance door's locks. But I was still worried for her to fall from the stairs in the darkness.
>
> (CE 1)

Other participants also indicated having safety concerns due to the physical characteristics of their home, such as the low height of their balcony's fences (CE 2), the type of windows they had in their houses (CE 4), or their door locking system (CE 3). Most safety concerns were caused by either unsuitable housing design or deficiencies in the safety features of the physical home environment.

Housing type

Housing type had a significant effect on increasing or decreasing CEs' psychological and physical health outcomes. Having multi-level spaces across two or three floors increased internal commuting distances and resulted in a waste of time for busy CEs. Having all the spaces contained within a single floor was favored by CEs since they could keep an eye on their care recipient and do their paid work at the same time. One of the participants expressed her disapproval of stairs (Figure 7.1), noting:

> I want to change it! It is really hard going up and down from these stairs every day for several times. . . . It really makes me tired. . . . Especially some days that my mom stays in her bedroom upstairs. . . . I need to go up and down thousands of times. . . . My knees have started to ache these past weeks. . . . I really wanted to live in a bigger house, which has everything in a single floor.
>
> (CE 3)

Living in a single house with stairs was specifically challenging given the need to perform tasks across a variety of rooms located on multiple levels, such as caring, working, and domestic management.

Size and layout

In addition to housing type, the size of the house and that of its internal spaces were significantly important for fulfilling CEs' physical environmental needs. Having a small house resulted in several negative consequences for CEs, such as a lack of working place for paid labor, which allowed for privacy, concentration, and personal space. All of the Iranian CEs previously had big houses in Iran, which made it more difficult for them to live in smaller houses. One participant explained her biggest spatial tension as:

> Not having space! Not having space! Deficiency in number of rooms, having lots of noises, too much traffic everywhere . . . because I could not afford a bigger place and I did not have the money. . . . I had a very nice home back in Iran and I miss it. . . . If you have a bigger house, you may have more space for yourself and less noise and more concentration.
>
> (CE 4)

Related to this, CEs did not have a dedicated personal space for themselves due to the scarcity of space. Given that women are in charge of almost everything in the Iranian household and are expected to be available to all family members at home, the CEs spoke about the lack of such a space:

> Maybe if I had a bigger house, I could have a space for my own and told everyone that this place is completely mine. I did not have such place, but my husband did. When he wanted to do his things, he picked his laptop and went to the basement to have his 'peace of mind'.
>
> (CE 4)

Figure 7.1 Stairs present spatial and temporal challenges

A larger house would allow CEs to not only better manage their multiple responsibilities but also provide some personal space for restoration, which would have undoubtedly resulted in enhanced well-being.

Furniture

Household furniture was either used as assistive equipment to ease CEs tensions or the source of tension itself. The hardship of caring responsibilities seemed to be doubled due to the inappropriate use of furniture. CEs paid work productivity was also negatively impacted by the quality of the furniture. One CE discussed the unsuitability of her working place as follows (Figure 7.2):

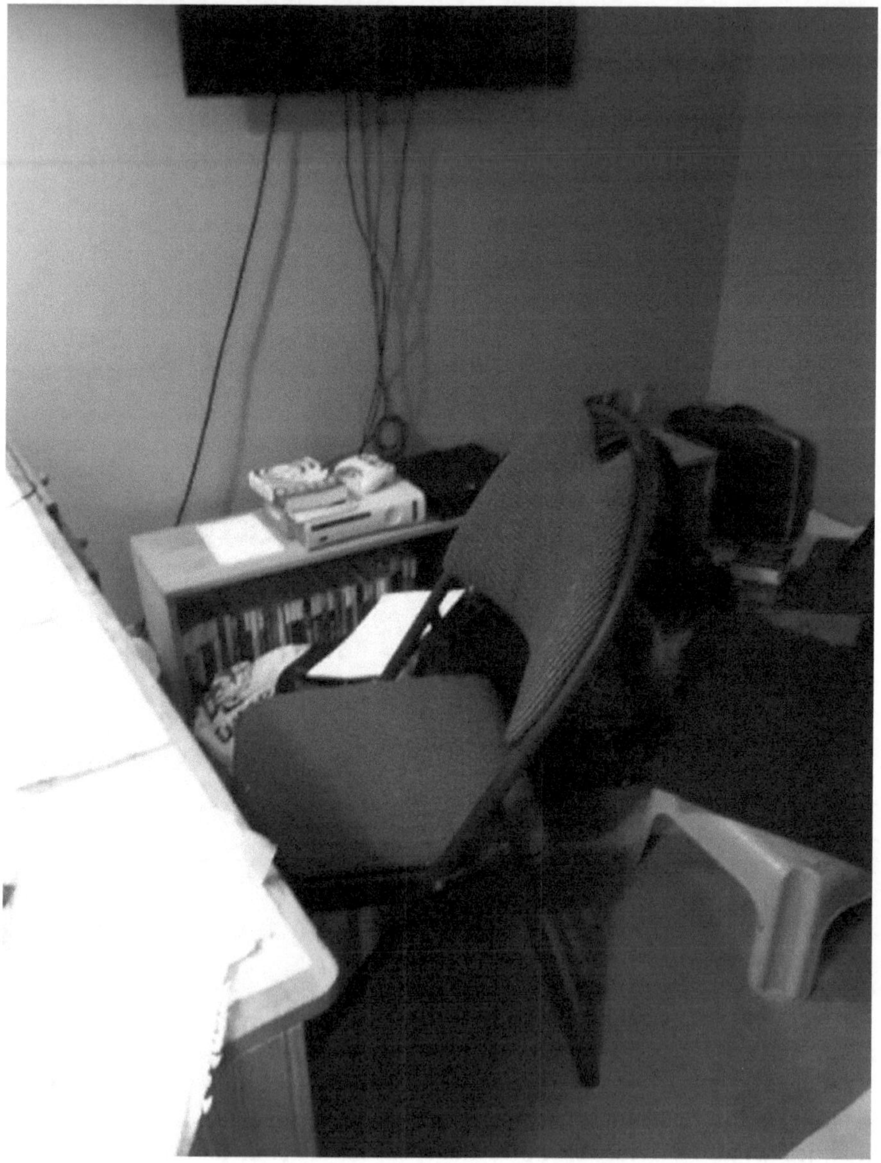

Figure 7.2 Unsuitable furniture

> Here we have two lamps on the table and a ranch chair which is not a computer chair. So, I need to put this chair back with the dining table when I am finished with my work. So you can see how it is messy. This is the condition which I study and do my work. This is my working place.
>
> (CE 4)

This CE often needed to work on the dining table due to marrying of her caring and domestic responsibilities. In the Iranian culture, all of the family members gather in the central space of the house in the evening and spend time together watching TV, eating, and chatting. Thus, the female carer who sat at the kitchen table near the living room could not concentrate on her work given the noisy environment.

Ambient features (light, view, noise)

The results suggest that the home's ambient features, such as lighting, views, and noise, have a significant effect on CEs' mental and physical health. Experiencing too much noise during working hours was reported to be extremely stressful. CEs were not able to finish their work as a result of the distraction caused by the noisy working environment. An insufficient amount of natural light and an unsatisfactory view had several negative consequences for CEs. One of the participants emphasized the importance of lighting quality and windows as follows:

> We did not have many windows in our previous house . . . for the view, you can create the view for yourself, but for the natural lighting, it was not complete. When it is dark in the house, you become depressed, especially when you are at home all the time.
>
> (CE 5)

Sufficient natural and artificial lighting was especially important for CEs' working space. CEs were most uncomfortable when bad lighting characterized their working space. A participant who was forced to work at the dining table complained about the quality of lighting:

> The lighting on the kitchen table was not good . . . because of that, I could not concentrate on my work.
>
> (CE 4)

Seeing natural elements and having nice views in the home environment decreased CEs stress and contributed to their mental well-being. This was especially the case for spaces where they spent most of their time, including the kitchen, living room, and working space. Iranian CEs praised Canada for its beautiful and more accessible natural views, compared to Iran. Visually connecting with nature was significantly important for them, given their limited time outside of their homes. One of the participants expressed her disappointment with the lack of a natural view in her work space (Figure 7.3):

Figure 7.3 A view of nature but on the wrong floor

> Well, for the place that I want to change, I love to see nature when I am working. But you know because my first floor is not next to the backyard, I don't have that view. My backyard is next to the basement. I really wanted my first floor to be next to the back yard. I cannot say how much I would have loved that.
>
> (CE 2)

In addition to quality lighting and natural view, nearby noise was reported to be a major source of tension for the CEs. Resting and having a good night's sleep is critical for CEs.

Privacy

Privacy is an essential need among all human beings and CEs are not an exception. CEs need private time, especially while doing paid work. Lack of privacy was a major spatial tension identified, resulting in negative psychological consequences. Most CEs required a personal place for themselves, in order to relax or work without any disturbance. At the same time, they got nervous if they could not see their care recipient. One participant explained privacy during working hours as follows:

> My working space is not that separate. Because it's out there. You can see that I am here working or studying, mom is there on the sofa. I want more privacy when I work, but I will be nervous if I cannot watch mom.
>
> (CE 3)

She showed the lack of privacy in her working space in a photo of her living room. Other carers had the same problem with either their working or personal space (Figure 7.4).

Figure 7.4 A working space with limited privacy in the living room

Structural components

Structural components refers to the architectural elements that form the structure of the building, such as walls, stairs, ramps, windows, doors, and floors. The unsuitable design, size, location, color, quantity, and safety of each of these elements play a significant role in increasing spatial tensions for CEs. For example, the color of the walls was noted to be important:

> I wanted the colour of my home to be a little bit friendlier . . . for example instead of the painful not washable garbage red colour, we had the sky blue. You know, when you have a brighter colour, it makes you feel better.
>
> (CE 4)

Sometimes having a wall separating two places from each other was a major source of tension and stress for carers since they were not able to constantly monitor their care recipient. The floor's color and material were also important for CEs, from both cleaning and caring standpoints. One of the participants stated:

> In our house, the floor has a red colour and it is very bad, since you can clearly see the dirt. I wanted to have a bright colour for the floor, so that it will appear cleaner. My mom always drops her food on the floor, so I need it to be easily cleaned.
>
> (CE 1)

A range of structural components were highlighted as important to minimize spatial tensions and improve quality of life.

Temporal tensions

Temporal tensions are defined as pressures and challenges faced by CEs in specific time periods, or over an extended period of time. These tensions were mostly in the form of different activities that CEs had to realize, and are organized into two main themes: monitoring and control, and daily activities.

Monitoring and control

As explained earlier, one of the most important caring responsibilities associated with the CEs was the monitoring and control of the care recipient. Monitoring and control were conducted either constantly throughout a long period of time or temporarily. Most carers stated that they needed to monitor and control their care recipient "all the time," noting that this task was very difficult and made CEs feel extremely stressed. For instance, one CE emphasized the need to take care of her mother who had Alzheimer's disease:

> You know, she is exactly like a one or two year old kid. You need to look after her all the time! . . . I always have that concern and stress in my head.

> And I am watching and looking out for my mom all the time . . . so the stress is always there.
>
> (CE 1)

Similarly, other CEs also expressed their concern regarding the consistent monitoring that was required "all the time."

Time involved in daily activities

Daily activities occupied a major portion of CEs' time during the day and included caring, paid work, and cooking. CEs often mentioned how limited their time is in a day because of the numerous daily activities they perform. One CE expressed her limited time for daily activities accordingly:

> I feel like in 24 hours, the hours are not enough for me to do all of my responsibilities. I really have this stress with me all the time. From the midnight and while I am doing the last things for my mom before going to bed, I am already thinking about the next morning and how I should manage all of the things I am supposed to do.
>
> (CE 2)

A great portion of CEs' daily activities were caring responsibilities, as described above. Some of these responsibilities were strictly time-dependent, such as providing medicine to the care recipient at specific times in a day. One participant took a photo of her mother's nightstand table, showing the many different medicines she needed to give her mom throughout specific times of the day (CE 3, Figure 7.5).

Another major daily activity was meal preparations. Although Iranian carer-employees lived in Canada for many years, they maintained the traditionally division of labor between men and women when it came to meal preparation. For example, all female CEs were responsible for preparing breakfast, lunch, and dinner on a daily basis, receiving no help or assistance in cooking and food preparation.

Discussion

Both spatial and temporal tensions were experienced by the Iranian CE sample, causing them to experience negative physical, psychological, and social impacts associated with stress and anxiety. Spatial and temporal tensions were determined in the micro-spaces of the home, as identified through self-captured photos using Photovoice. Spatial tensions were related to safety, housing type, size and layout, furniture, ambient features, lack of privacy, and structural components of the home environment, while temporal tensions were related to the monitoring and control of the care recipient, and those associated with daily activities, such as meal preparation. Poor home design was realized to be the primary cause of these tensions.

Figure 7.5 Various medications be given at different times of the day

The research results suggest that immigrant carer-employees are not only working hard to uphold the responsibility to care for aging family members in a new country, but also simultaneously caring for other dependents, doing the bulk of the domestic work, and choosing to work from home in order to meet their multiple demands while contributing to the comparatively higher costs of living in Canada. The data suggest that gender inequalities remain deeply entrenched in Iranian immigrant society, with women having full and primary responsibility for unpaid care and domestic work.

Although significant progress has been made in advancing gender equality, through landmark agreements such as the *Beijing Declaration and Platform for Action* and the Convention on the Elimination of All Forms of Discrimination against Women (CEDAW), much work is needed to move SDG 5.4.1 given the culturally entrenched role of unpaid care and domestic work lying on women's shoulders in most ethnic groups, immigrant or otherwise (UN Women, 2015). In July 2010, the United Nations General Assembly created UN Women as the entity for gender equality and women's empowerment. UN Women offices across the globe have been engaged in measuring SDG 5.4.1, in order to track all progress being made over time, toward 2030 (UN Women, 2015). The focus for many countries has been the addition of time use questions, specific to unpaid care and domestic work, added to national census and labor force surveys, for example. Such quantitative time-use data should best be complemented with spatial data in order to fully understand the impact of unpaid care and domestic work on carers. Further, mixed methods and qualitative approaches provide rich data characterized with a depth of understanding of the phenomenon of the experience of unpaid care and domestic work, the majority of which is done in the home. Another solution for decreasing immigrant women CE's tensions is to increase the awareness of the division of labor within their households. Working toward an equal division of labor across both sexes, where both men and women are equally sharing family care and domestic work, is essential for women's empowerment and health. Conducting research like the one described in this chapter can assist in reaching this goal, reflected in all participants noting that the Photovoice process was "informative," in that it made them more aware of their needs, goals, and vision for their life. This Photovoice study on Iranian immigrant carer-employees suggests that there are both spatial and temporal tensions at play in balancing care, domestic work, and paid work in the micro-spaces of the home. It suggests that stress and anxiety are commonplace among CEs, having negative physical, mental, and psychological impacts on their health and well-being.

Acknowledgments

Our greatest thanks and appreciation to the Iranian CEs for participating in this study.

Declaration of interest statement

No potential conflict of interest was reported by the authors.

Funding

This study was funded via Canadian Institute of Health Research (CIHR) Research Chair in Gender and Health, addressing Caregiver-Friendly Workplace Policies [MOP-60484].

References

Braun, V., & Clarke, V. (2006), 'Using thematic analysis in psychology', *Qualitative Research in Psychology*, 3(2), pp. 77–101. doi: 10.1191/1478088706qp063oa

Dardas, A., Williams, A., Kitchen, P., & Wang, L. (2019), 'Assisted-transport caregiving and its impact towards carer-employees', *Journal of Gerontological Social Work*, 62(4), pp. 475–497. https://doi.org/10.1080/01634372.2019.1596184

Evans, J., & Jones, P. (2011), 'The walking interview: Methodology, mobility and place', *Applied Geography*, 31(2), pp. 849–858. doi: 10.1016/j.apgeog.2010.09.005

Hay, I. (2010), *Qualitative research methods in human geography*, 3rd edn. London, UK: Oxford University Press.

Ilagan, I., Akbari, Z., Sethi, B., & Williams, A. (2020), 'Use of photovoice methods in research on informal caring: A scoping review of the literature', *Journal of Human Health Research*. https://doi.org/10.14302/issn.2576-9383.jhhr-20-3573

Kwan, M. (2000), 'Interactive geovisualization of activity-travel patterns using three-dimensional geographical information systems: A methodological exploration with a large data set', *Transportation Research Part C*, 8(6), pp. 185–203.

Ramesh, S., Ireson, R., & Williams, A. (2017), 'International synthesis and case study examination of promising caregiver-friendly workplaces', *Social Science & Medicine*, 177, pp. 52–60. doi: 10.1016/j.socscimed.2017.01.052

Spitzer, D., Neufeld, A., Harrison, M., Hughes, K., & Stewart, M. (2003), 'Caregiving in transnational context', *Gender & Society*, 17(2), pp. 267–286. doi: 10.1177/0891243202250832

United Nations (2020), *Sustainability Development Goals: Department of Economic and Social Affairs, Statistics Division* [Online]. Available at: https://unstats.un.org/sdgs/metadata?Text=&Goal=5&Target=5.4

UN Women (United Nations Entity for Gender Equality and the Empowerment of Women) (2015), *Beijing Declaration and Platform for Action Convention on the Elimination of All Forms of Discrimination against Women* [Online]. Available at: www.un.org/en/events/pastevents/pdfs/Beijing_Declaration_and_Platform_for_Action.pdf

Wang, C., & Burris, M.A. (1997), 'Photovoice: Concept, methodology, and use for participatory needs assessment', *Health, Educucation & Behaviour*, 24(3), pp. 369–387.

8 Resource insecurity and gendered inequalities in health

A challenge to sustainable livelihood

Godfred O. Boateng

Introduction

In September 2000, the governments of 189 countries adopted the United Nations Millennium Development Goals and resolved to "spare no effort to free our fellow men, women, and children from the abject and dehumanizing conditions of extreme poverty, to which more than a billion of them are currently subjected" (UN Millennium Declaration, 2000). Today, over 700 million of the world's population still live in extreme poverty, with the majority being females (World Data Lab, 2021). Resource insecurity is symptomatic of broader economic disadvantages and social inequalities faced by the poor (Boateng *et al.*, 2021) and quite distinct from resource scarcity (Wutich and Brewis, 2014), although the latter is considered a component of the former.

Resource scarcity has been examined from minimalist, moderate, and maximalist perspectives. Based on the minimalist account, resource scarcity is concerned with the availability of natural resources needed to satisfy basic human needs for food, shelter, and energy. The moderate perspective examines resource scarcity as concerning the availability of resources to satisfy consumption at current or higher levels. The maximalist perspective considers resource scarcity in terms of the actual demand of both human and non-human species exceeding supply (Matthew, 2008). In sum, resource scarcity can be defined as the shortage of resources influenced by a decline in availability, quantity, quality, or efficiency that does not meet current or increased demand, and has implications for sustainability and survival of human and non-human species. While this definition of resource scarcity appears to be holistic, the definition for resource insecurity is more encompassing.

This chapter will adapt Hadley and Wutich's (2009) definition of resource insecurity with some modifications. For the purposes of this chapter, resource insecurity is defined as a multifaceted concept that encompasses resource scarcity, resource access, and lifestyle concerns that have the potential to impact the survival and sustainability of the human species. The complex and multidimensional nature of resource insecurity makes it a broad concept, requiring the need to identify its scope and scale in this chapter. A previous reflection on resource insecurity focused on food and water shortage, the causes of resource insecurity

DOI: 10.4324/9780367743918-10

at the community level, exploring the coping responses at the household level, and examining its effect on the emotional well-being and mental health at the individual level (Wutich and Brewis, 2014). While that chapter sought to develop a broader theory of resource insecurity at the level of the individual and the household, as well as examine the emotional and mental health effects, it was deficient in the composition of resource insecurity and did not draw adequate attention to the extent to which such insecurities exacerbated the health equity gap. This chapter has a different focus, emphasizing the multiple elements that make up resource insecurity, how it threatens health and sustainable livelihoods, and exacerbates gendered inequalities.

Accordingly, resource insecurity will be examined as a holistic concept made up of a triad, consisting of food, energy, and water insecurity at the household and individual levels. Resource insecurity will not be examined in isolation, as it is important to explore the causes and consequences of resource insecurity. This warrants a synthesis of the literature on some of the factors that influence resource insecurity, as well as consequences. The consequences of resource insecurity will be examined through four pathways, including disease, nutrition, economic, and psychosocial consequences, and how it perpetuates gendered inequalities. Using a transdisciplinary approach, I will examine the co-occurrence of food, water, and energy insecurity, and the multiplicative effect it has on amplifying gender inequality.

Resource insecurity and gender inequality

Resource insecurity in the form of food, water, and energy can be differentiated into unique components. The Food and Agriculture Organization (FAO) of the United Nations states that food security exists "when all people, at all times, have physical and economic access to sufficient, safe and nutritious food to meet their dietary needs and food preferences for an active and healthy life" (FAO, 1996). This suggests that the absence of any of these components is indicative of food insecurity. Along these lines, the US Department of Agriculture defines food insecurity as "a household-level economic and social condition of limited or uncertain access to adequate food" (US Department of Agriculture, Economic Research Service, 2006). Analogous to the definition of food insecurity, Jepson *et al.* (2017), after a synthesis of existing conceptualizations, defined water security as the ability to access and benefit from affordable, adequate, reliable, and safe water for well-being and a healthy life. Here again, the condition where at least one of these components (affordability, reliability, adequacy, and safety) is significantly reduced, or becomes unattainable and threatens or jeopardizes well-being, provides a description of water insecurity (Jepson *et al.*, 2017). A parallel definition has been given to energy insecurity as the lack of access to adequate, affordable, reliable, acceptable, and clean sources of energy for a healthy and sustainable livelihood (Boateng *et al.*, 2020). Taken together, these definitions highlight key components of resource insecurity to include the lack of access to adequate, affordable, reliable, acceptable, and safe sources of either food, water, and/or energy for an active, healthy, and sustainable livelihood. Access to all three of these components would imply resource security. While each of these components have been examined at the international, regional, and national levels, the impact of this triad, particularly food and water insecurity at the household and individual levels, have gained the attention of academics and policy makers, with energy insecurity considered a burgeoning field of study. All three are inextricably linked to each other and have the capacity to amplify the effects of each other via a shortage in one (Boateng *et al.*, 2020). Resource insecurity in its current characterization is at the heart of human development and is intricately tied to the attainment of several of the sustainable development goals (SDG) including the reduction of poverty (goal 1); ending hunger (goal 2); ensuring healthy lives and promoting well-being (goal 3); achieving gender equality and the empowerment of women and girls (goal 5); ensuring access to affordable, reliable, and sustainable energy (goal 7); reducing inequalities (goal 10); making human settlements safe resilient and sustainable (goal 11); ensuring sustainable and production patterns (goal 12); and combating climate change (goal 13) (United Nations Department of Economic and Social Affairs, 2015).

Current estimates before COVID-19 show that nearly 690 million people, or 8.9% of the world's population, were food insecure or hungry. More than 250 million of the food insecure live in Africa, where the number of undernourished people is growing faster than the rest of the world (FAO, 2020b). Globally, the prevalence of moderate or severe food insecurity is higher among women than among men (FAO, 2020b). Equally, water insecurity estimates show that about 3.2 billion people live in agricultural areas with high to very high-water shortages or scarcity, and 1.2 billion of this number (roughly one-sixth of the world's population) live in severely water-constrained agricultural areas. The most at risk are the poorest and most vulnerable groups, including women (FAO, 2020a). Water shortages and scarcity have been found to jeopardize the environment that is necessary to enable and ensure access to food for millions of hungry people in many parts of the world (FAO, 2020a).

The estimates of energy insecurity are no different. The 2017 international Energy Agency report on transitions from poverty to prosperity shows that one in five people in the world lack access to electricity and about 2.8 billion lack access to clean cooking (International Energy Agency, 2017). Future projections show that by 2030, about 600 out of the 674 million people in SSA will still be without access to electricity (International Energy Agency, 2017). Here again, the impact is particularly severe for women and girls, who are typically responsible for securing and using energy resources (Boateng *et al.*, 2020).

The current statistics of resource insecurity are indicative of a growing problem, as the prevalence of each of the three components continues to increase, with the worst affected living in SSA. The situation is estimated to be worsened by the COVID-19 pandemic, as several thousands of people around the world plunge deeper into poverty. Spatial inequalities in food, energy, and water access have resulted in differential health, disease, psychosocial, and nutritional effects for several populations. This is particularly the case for women and children across space and time, leading to social and health disparities. While there is ample evidence of gender inequality in resource access (Kassie, Ndiritu and Stage, 2014; Wutich and Brewis, 2014; Tibesigwa and Visser, 2016; Quinonez, de Sousa and Figueroa, 2019; Botreau and Cohen, 2020), few studies have explored the relationship between resource insecurity and gender inequality.

Gender equality is not only a fundamental human right but also a necessary foundation for a peaceful, prosperous, and sustainable world. While some progress has been made over the last few decades, there are pervasive challenges which have made stagnate social progress, and have kept women bearing the severe consequences of resource insecurity. Globally, women and girls perform a disproportionate share of unpaid domestic work (United Nations Department of Economic and Social Affairs, 2015). They are responsible for ensuring there is sufficient food, energy, and water resources at home, in spite of the disproportionate time they spend in acquiring and using these resources. This gendered division of labor is influenced by socioeconomic and political processes that structure hierarchical power relations between men and women. These factors determine who has what (material and other assets), who does what (division of labor between market and reproductive labor), who decides what (political participation and law), and who is valued for what (social norms, ideologies) (George *et al.*, 2020). These factors ensure the practice and perpetuity of gender inequality. In fact, gender norms govern what is considered acceptable for men and women. These norms legitimize patriarchy and conceal its unfairness; it is also what legitimizes the differential roles of men and women (George *et al.*, 2020). This creates a situation where women begin to experience inequality at birth, following them all through their lives.

Socio-economic and political processes also influence who has what. Structures, such as discrimination in land, property, inheritance laws, access to low-cost credit, and access to other economic opportunities, disproportionately affects women who are responsible for the management of food, energy, and water resources in the household (George *et al.*, 2020). Gender norms also influence who does what. It determines the role of women at the household and community level. It supports early marriage and positions women and girls as responsible for the household and care work, which often limits their completion of education, participation in the labor force, and upward economic mobility. This denies women the autonomy and the capacity to engage in decision-making at the household level. In 18 countries, husbands can legally prevent their wives from working; in 39 countries, daughters and sons do not have equal inheritance rights; and 49 countries lack laws protecting women from domestic violence (United Nations Department of Economic and Social Affairs, 2015). In sum, these factors create a situation where women and girls are disproportionately impacted by resource insecurity, which negatively impacts their well-being and sustainable livelihood. These differential effects are set to increase as resource insecurity grows with accelerated planetary climate change, population increase, protracted weather conditions, biodiversity loss, mass extinctions of plants and animals, destruction of ecosystems and natural habitats, and increase in extreme poverty due to the COVID-19 pandemic (Barnosky *et al.*, 2012; Campbell *et al.*, 2017; Hasegawa *et al.*, 2018). Beyond the impact of COVID-19, the capacity of countries, especially, those in SSA to recover from the reversal in the progress made on the SDGs, is worrisome. As extreme poverty deepens, the access to resources such as food, water, and energy become protracted, resulting in widening of the gender inequality gap.

The remainder of this chapter explores the current state of knowledge about the different pathways by which resource insecurity impacts on disease, nutrition, economic, and psychosocial well-being of males and females in most low- and middle-income countries.

I cannot exhaustively review the extant knowledge about these pathways across the different disciplines. Rather, I focus on four main domains for which the evidence is undisputable. Next, I will discuss the syndemic effect of resource insecurity in perpetuating the cycle of poverty and the gendered inequalities in health. Finally, I will identify possible strategies that could be used to mitigate these effects.

Resource insecurity and differential health effects

The literature on resource insecurity has provided ample evidence on the deleterious consequences of food, energy, and water insecurity at the individual and household levels. However, fewer studies have explored how this insecurity exacerbates and perpetuates gender inequality. This section examines the evidence on the disease, nutrition, economic and psychosocial consequences of food, energy, and water insecurity, and how it differs across sex (Table 8.1).

Disease pathway

Resource insecurity, whether in the form of food, energy, or water insecurity, has deleterious health consequences on individuals and households, with women and girls experiencing the worst effects. Food insecurity is a major cause of malnutrition in low-income countries, with maternal and child undernutrition contributing to more than 10% of the world's disease burden (Alaimo, Chilton and Jones, 2020). Among children under five years, a recent review of food insecurity and nutrition indicators revealed the positive significant association between food insecurity and stunting, child wasting, low birthweight, and anemia (Maitra and Food and Agriculture Organization of the United Nations, 2018). Among adults, food insecure households are more likely to be overweight and obese due to the decreased intake of fruits, vegetables and fiber, together with an increased intake of energy dense foods, including foods that are rich in fat and sugar (Shariff and Khor, 2005; Popkin, 2011; Castillo *et al.*, 2012). However, the prevalence of food insecurity induced obesity is greater among women and adolescents, and not men (Ivers and Cullen, 2011). This has implications for an increased risk of diet-sensitive chronic diseases, including hypertension, dyslipidemia, diabetes, and cardiovascular diseases (Gundersen and Ziliak, 2015). Food insecurity during pregnancy also increases the risk of weight gain and gestational diabetes mellitus

Table 8.1 Resource insecurity and differential health and economic outcomes

Domains	*Resource insecurity*	*Consequences*
Disease	*Food Insecurity*	Cardio-metabolic disease; type 2 diabetes; dyslipidemia; hypertension; frequent sickness of children, anemia, stunting, wasting, low birth weight deliveries
	Energy Insecurity	Chronic disease; cardio-metabolic disease; hypertension; acute lower respiratory infection; anemia; myopia; chronic obstructive pulmonary disease; chronic bronchitis, ischemic heart disease; hypothermia; hyperthermia; pneumonia; Gastrointestinal disease; diarrhea; hepatitis A; bodily injuries
	Water Insecurity	Gastrointestinal diseases; water borne diseases; cholera; dysentery; diarrhea; typhoid and paratyphoid enteric fevers; hepatitis A, malnutrition; anemia; hemorrhage; food poisoning
Nutrition	*Food Insecurity*	Malnutrition; stunting; wasting; severe caloric deficiency; poor dietary diversity; low intake of micronutrients; starvation; undernutrition; unbalanced diet; unhealthy dietary patterns; water insecurity
	Energy Insecurity	Low food production; lower dietary diversity; food poisoning; water poisoning; fewer food choices; energy expenditure
	Water Insecurity	Poor food quality; lower caloric intake; energy expenditure; lower dietary diversity; lack of nutrient rich foods; increased food insecurity; dehydration
Economic	*Food Insecurity*	Impairment or delays in cognitive development; poor academic performance; lower productivity; unemployment; low production levels; deficiency in human capital formation; human capital deficits; decreased life time earnings; school readiness, lower educational attainment; high rates of school absenteeism and repeating grades; lower life-time earnings; low work capacity; increased out of pocket payments for hospitalization
	Energy Insecurity	Low food production; income generation activities; household income; school absenteeism; poor academic performance; increased energy expenditures; decrease in productivity
	Water Insecurity	Decreased income generation activities; increase water expenditures; increased expenditures on health; increased school absenteeism; lateness to school; human capital deficits; food insecurity; decreased productivity
Psychosocial	*Food Insecurity*	Anxiety and depressive symptoms; mood disorders; anxiety disorders; stress; emotional health; elevated mental distress; impaired mother to child interactions; impaired mother to child attachment; child neglect and abuse; low self-esteem.
	Energy Insecurity	Stress; depression; anxiety; social exclusion; feelings of exhaustion; shame; low self-esteem; reduced maternal-child interactions
	Water Insecurity	Stress; depression; worry, anger, reduced maternal-child interactions; deficiency in infant cognitive development; intimate-partner violence

(Laraia, Siega-Riz and Gundersen, 2010), as well as low birth weight. The persistence of food insecurity means the goal of ending all forms of malnutrition and ending hunger as well as ensuring access by all people, in particular the poor and people in vulnerable situations by 2030 has become illusive. Thus, vulnerable populations such as women and girls would experience the severe effects of food insecurity.

Equally, water insecurity contributes to the spread of infections and waterborne diseases. The WHO's recent report on drinking water shows that water and poor sanitation are associated with the transmission of diseases such as cholera, diarrhea, dysentery, hepatitis A, and typhoid (Tarrass and Benjelloun, 2012; WHO, 2019). The lack of access to safe water means households are compelled to use surface water, which may contain *E. coli* and other pathogens that increase the risk of gastrointestinal diseases (Collins *et al.*, 2019). About 829,000 people are estimated to die each year from diarrhea as a result of unsafe drinking water, sanitation, and hand hygiene (WHO, 2019). Diarrheal diseases are a major contributor to malnutrition and death. Diarrheal diseases stem from anorexia, reduced absorptive function, and mucosal damage, in addition to nutrient exhaustion, the latter of which is associated with each episode of diarrhea (Ferdous *et al.*, 2013). Pregnant women who have to carry heavy water containers may also increase the risk of uterine prolapse and hemorrhage (Gjerde *et al.*, 2017). These effects say little about the progress made by most SSA countries on achieving universal and equitable access to safe and affordable drinking water for all by 2030. Without access to safe and affordable drinking water, it is difficult to imagine what progress could be made in achieving equity in sanitation and hygiene for women and girls.

Comparably, the burgeoning literature on household energy insecurity suggests both direct and indirect health consequences. Particularly, energy insecurity in most low-income countries has necessitated the reliance on unprocessed solid fuels for cooking, leading to household indoor air pollution and an increase in accidental deaths (Dherani *et al.*, 2008; Wilkinson *et al.*, 2009; WHO, 2014). Through indoor air pollution, households' members are exposed to high concentrations of particulate matter, gases, and other pollutants. This is due to coal burning in open fires, or low efficiency stoves, both which result in acute lower respiratory infections, chronic obstructive pulmonary, and ischemic heart diseases (Cook *et al.*, 2008; Wilkinson *et al.*, 2009; Boateng *et al.*, 2020). Women and girls are at a particularly high risk of carbon monoxide, sulfur and nitrous oxides, and hydrocarbon poisoning. This may be attributed to the gendered roles specific to household responsibilities, requiring women to spend more time cooking indoors with their girls (Boateng *et al.*, 2020). Also, the lack of energy resources has resulted in the use of salt for the preservation of foods, which has also led to an increase in chronic diseases, particularly, hypertension. Women sleep thirsty so their children will have water to drink. These consequences do not match the goals of SDG 7. Any progress in eliminating the deleterious effects experienced mostly by women and girls in the area of energy insecurity might require a proactive measure to ensure universal access to affordable, reliable, and modern energy services.

By examining all three components of resource insecurity, it is obvious that any deficiencies in the achievement of key targets related to SDG 2, 6, and 7 may have a greater effect in the reduction of poverty (Goal 2) and ensuring health lives (Goal 3). With a reversal in the progress made on achieving resource security, resource insecurity manifesting in the form of food, energy, and water insecurity will continue to have disproportionate consequences on women and girls than men.

Nutrition pathway

Resource insecurity in its current description has been associated with nutritional consequences among low-income earners. In most low-income countries, food insecurity is associated with negative nutritional outcomes and it is experienced differently by women and men. Generally, food insecurity is associated with severe caloric deficiency, poor dietary diversity, low intake of micronutrients, starvation, unbalanced diet, and unhealthy dietary patterns (Sen, 1977). Among the elderly, food insecurity has also been found to be significantly associated with lower intakes of energy, protein, carbohydrate, saturated fat, niacin, riboflavin, magnesium, iron, and zinc (Lee and Frongillo, 2001). Among women, food insecurity is associated with changes in body composition during lactation. A longitudinal study in Uganda showed that food insecurity was associated with lower body weight at 6, 9, and 12 months postpartum (Widen *et al.*, 2017). A related study in Kenya showed that food insecurity was inversely associated with arm muscle area and mid-upper arm circumference among postpartum women (Widen *et al.*, 2019). More recently, greater food insecurity was found to be significantly associated with lower breast milk intake among infants in western Kenya (Miller *et al.*, 2019).

Water is life, as it represents a critical nutrient for survival. However, the absence of water has dire consequences for households, particularly for women and children. In many low-income settings, water insecurity has been associated with dehydration in adults and children (Popkin, D'Anci and Rosenberg, 2010; Rosinger, 2018). The worst form of dehydration was reported by Krumdieck *et al.* (2016), who found women sleeping thirsty in Kenya due to lack of access to water. Again, among adults and children, dehydration has also been associated with disruptions in mood and cognitive functioning. Inadequate intake of fluids is associated with increased prevalence of constipation (Popkin, D'Anci and Rosenberg, 2010). Water insecurity is also reported to decrease the quality, quantity, and the diversity of food consumed, and is shown to increase energy expenditure for women (Collins *et al.*, 2019). Indeed, in a longitudinal study of postpartum women in Kenya, water insecurity was established as an important determinant of food insecurity (Boateng, Workman *et al.*, 2020).

The nutritional consequences of energy insecurity for women and children are quite significant. Energy is critical for food production, improving the quality and safety of food and water, and increasing the variety of food accessible to households. Homes lacking access to energy are likely to have fewer food choices, greater food-borne illness due to undercooking, and more food poisoning due to improper refrigeration (Hernández, 2016; Boateng *et al*., 2020). In fact, energy insecurity can be considered a determinant of food insecurity. Beyond the household, energy is required for food production and processing, particularly, in the irrigation process amidst extreme temperatures. The absence of such energy means low food production, which has implications for child malnutrition or undernutrition.

The nutrition effects of resource insecurity challenge the possibility of achieving four (1, 2, 3, 6) of the 17 sustainable development goals. For instance, the lack of micronutrients suggests ending all forms of malnutrition, whether stunting or wasting in children under five years of age by 2025 (Goal 2) is far from reality. Also, with protracted weather conditions and the drying of water bodies, women have had to sleep thirsty so that their children would have water to drink due to the lack of access to sufficient and safe water in Madagascar. This global inequality in water access begs the question whether universal and equitable access to safe and affordable drinking water for all (Goal 6) would be achieved in time to ameliorate the physical, economic, and psychological burden experienced by women.

Economic pathway

The economic consequences of food, water, and energy insecurity are significant and disproportionately impact the livelihoods of women and children, with lifelong consequences.

Starting with children, food insecurity has been found to prevent children from attaining their full potential (Cook and Jeng, 2009). Thus, children who are consistently hungry may experience development impairments which limit their physical, intellectual, and emotional development (Cook and Jeng, 2009). This has other related effects, such as delays in school readiness and limited learning or academic achievement, which in turn impacts their educational attainment, lifetime earnings, and determines their standard of living (Cook and Jeng, 2009). One of the key mechanisms to this poor performance is absenteeism. The evidence in a longitudinal study in Ethiopia and a cross-sectional study in Ghana (Belachew *et al*., 2011; Baiden *et al*., 2020), respectively, has shown that children from food insecure households are more likely to be absent from school with consequences on their academic performance (Shankar, Chung and Frank, 2017). This problem is not only limited to Low- and Middle-Income Countries (LMICs) but also prevalent among children from food insecure households in high-income countries (Shankar, Chung and Frank, 2017). Among adults, workers who experienced hunger as children are not well prepared physically, mentally, emotionally, or socially to perform effectively in the workforce (Cook and Jeng, 2009). Also, hunger has been found to increase the healthcare costs of families. A society that privileges males over females might put enough resources into supporting the male child. Furthermore, the experience of hunger by children has been associated with greater absenteeism, presenteeism, and turnover in the work environment, which is costly to employers. As women are predominantly responsible for care roles in the families, sick children often lead to parent–employee absenteeism (Cook and Jeng, 2009), which mostly have adverse consequences on career development and mobility of women in the workplace.

The economic effects of water and energy insecurity are somewhat related. With climatic changes, drought, and the short spells of rainfall, women and girls have to walk long distances to collect water and energy resources. Women in Kenya spent between two and 10 hours a week collecting water for the household (Boateng *et al.*, 2018). Women in India spend approximately 374 hours every year, and up to 20 or more hours per week, collecting firewood in India; in fact, they spend four hours every day cooking when using traditional stoves (Clean Cooking Alliance, 2015). This takes away from the productive activity of these women and girls. For women, it means a shortfall in their economic activity and their capacity to generate income, which drives gender inequality within the household (Wutich and Brewis, 2014). For girls, it leads to lateness to school and, in some dire cases, absenteeism from school several times in a semester (Wutich and Brewis, 2014; Collins *et al.*, 2019). This has the potential to impact: the academic performance of these girls, their ability to pursue further education, and their career prospects in the long term. Consequently, a perpetual cycle of poverty and gender inequality at the household level is created, which raises the importance of achieving the SDGs that aim to end poverty for all and empower women and girls (Goals 1 and 5). It also calls for greater attention to be paid SDG 10, specifically, on the empowerment and promotion of social, economic, and political inclusion of all, in this case, women and girls. In addition, SDG 10 calls for ensuring equal opportunity and reducing inequalities of outcomes. These targets will trigger efforts toward the achievement of full and productive employment and decent work for all women and equal pay for work of equal value (Goal 8, target 5).

Psychosocial pathway

There is ample evidence to suggest that resource insecurity has deleterious psychological consequences on households and their members. Several studies, including systematic reviews and meta-analyses, showed that food insecurity was associated with depression, stress, and poor emotional health (Bruening, Dinour and Chavez, 2017; Perkins *et al.*, 2018; Pourmotabbed *et al.*, 2020). Female-headed households have frequently been found to be more food insecure than male-headed households (Jung *et al.*, 2017); however, a study of food insecurity and mental health in 149 countries using Gallup polls shows that the psychological effects were similar for men and women (Jones, 2017). Nonetheless, women in this study scored more poorly on the mental health indices than men.

The psychosocial effects of water and energy insecurity are no different. Several studies have provided evidence for the effect that water insecurity has on emotional distress, psychological distress, psycho-emotional, psychosocial stress, and other forms of mental health (Wutich and Ragsdale, 2008; Aihara *et al.*, 2015; Stevenson *et al.*, 2016; Workman and Ureksoy, 2017a). Other psychosocial effects of water insecurity, based on the lived experiences of women in Kenya, consisted of anxiety, worry, shame, anger, and fear (Collins *et al.*, 2019). The women reported feeling ashamed when they were unable to maintain the hygiene of household members, or provide water as a gesture of hospitality (Collins *et al.*, 2019). In a comparative study between the experiences of males and females, Wutich (2009) found that women were more likely than men to be burdened with everyday water responsibilities; however, there were no significant differences between the experiences of males and females during household water emergencies and reports of worry, anger, and annoyance with family members (Wutich, 2009). A related study by Tsai et al. showed important intra-household gender differences in perceptions of water insecurity. Comparatively, women reported each aspect of water insecurity as being more severe than men (Tsai *et al.*, 2016). While these findings may suggest the existence of gender inequality in the effects of water insecurity, it is important to affirm that the disproportionate emphasis on women may explain the current results. Analogous to food and water insecurity, recent evidence from Ghana and Nigeria shows that women experience stress and anxiety, and may sometimes get depressed from carrying the disproportionate burden of securing adequate energy resources for the household. Participants reported experiencing more specific adverse consequences, including worry, shame, frustration, stigma, and quarrels (Boateng *et al.*, 2020). These psychological effects highlight the inextricably nature of resource insecurity in its effect and the urgent need to reduce poverty (Goal 1), end hunger in all forms (Goal 2), ensure the availability and equitable management of water resources (6), and make sustainable and modern energy affordable and available to all (Goal 7).

Taken together, resource insecurity has deleterious consequences in disease, nutritional, economic, and psychosocial pathways, in ways that widen the health gap between males and females to a greater extent. But most critically, it highlights the urgent need to address at least 10 of the 17 sustainable goals in SSA, without which the gender gap in access to food, water and energy would only deepen, with deleterious consequences on women's health.

Individual and syndemic effects of resource insecurity

Syndemic theory was introduced by anthropologists to explain the synergistic interaction of two or more co-existing diseases that amplifies excess burden of disease (Singer and Clair, 2003). Much earlier, Milstein (2001) described syndemics to occur when health-related problems cluster by person, place, or time. The problems, along with the reasons for their clustering, define syndemic and differentiate one from another. To circumvent a syndemic, Milstein argues that one must prevent or control not only each disease but also the forces that tie those diseases together. Biologically, the syndemic of diseases have been found to account for excess mortality (Singer and Clair, 2003). Following these explanations, Singer and Clair, in conceptualizing syndemic within the social context, argue that syndemic is "a set of intertwined and mutually enhancing epidemics involving disease interactions at the biological level that develop and are sustained a in a community because of harmful social conditions and injurious social connects" (Singer and Clair, 2003, p. 429). Using this framework, it is possible to see the syndemic relationships between food, energy, and water insecurity amplifying the associated health consequences and widening the health equity gap between males and females. Beyond this understanding, it has the potential to support policy makers and program implementers in improving population health (Tsai, 2018). Several studies have attempted to examine the syndemic relationships between the different components of this trifecta; however, none has explored the three together. A number of qualitative studies have provided evidence for the co-occurrence of food and water insecurity amplifying health outcomes. Workman and Ureksoy (2017b) in a study in Lesotho found that water insecurity, food insecurity and changing household demographics, likely resulting from HIV/AIDS epidemic, were all associated with increased anxiety and depression. The co-occurrence of food and water insecurity has quantitatively been associated with an increase in depression symptomatology (Boateng, Workman *et al.*, 2020) and undernutrition (Brewis *et al.*, 2020). From these studies, it is evident that the individual effects of food and water insecurity get compounded when they occur together. Thus, the co-occurrence of food, water, and energy insecurity would have a greater adverse effect on the household, and more severely on women and girls. The underlying force causing this insecurity is poverty. Households deep in poverty are more likely to experience food, water, and energy insecurity concurrently. This also means the effects as explained earlier, whether disease, nutrition, economic, or psychosocial, are compounded for women and girls who are disproportionately affected. Beyond poverty, other structural factors, including gender ideology, have a way of complicating the effects of food, water, and energy insecurity. For example, the gendered effects of resource insecurity lead to worst health and economic conditions for women; this, in turn, limits their access to resources, creating a cycle of poverty and perpetual gender inequality. In sum, the underlying force – poverty – is key to addressing resource insecurity, enhancing sustainable livelihoods, and reducing the health equity gap. Hence, ending poverty in all its forms everywhere in this context is particularly tied to ending hunger, and achieving food security and improved nutrition for women and girls (Goal 2); ensuring healthy lives and the promotion of well-being for women and girls at all ages (Goal 3); ensuring inclusive and equitable quality education for all (Goal 4); achieving gender equality and the empowerment of women and girls (Goal 5); achieving universal and equitable access to safe and affordable drinking water for all (Goal 6, target 1); ensuring universal access to affordable, reliable, and modern energy services (Goal 7, target 1); achieving full and reproductive employment and decent work for all women and men, including for young people and persons with disabilities, and equal pay for work of equal values (Goal 8, target 5); and ensuring equal opportunity and a reduction of inequalities in income (Goal 10, target 3) for women. Thus, the achievement of two or more of these goals would significantly reduce the health equity gap.

Conclusion

This chapter provides a synthesis of peer-reviewed articles to show how resource insecurity, in the form of food, water, and energy insecurity, influences gender inequalities in health. Each component has been discussed as having idiosyncratic effects on both households and individual members. Beyond these individual effects, resource insecurity using this triad has a synergistic effect on poverty. It is important to note that while each dimension may have an effect, the presence of either two or all three may worsen the experiences of those most disadvantaged and at risk. This amplification of effects therefore worsens the experiences of women, whether one considers the consequences (disease, nutrition, economic and/or psychosocial) of food, water, and energy insecurity independently or together. This makes the risk of morbidity and mortality, while precarious even with food insecurity only, much more complicated when experienced in the presence of water and energy insecurity. Interventions that target just one component of resource insecurity may not be efficient as all three facets are inextricably related with each other. To ensure sustainable livelihoods and reduce the gender inequality gap created by resource insecurity, it is important to examine the structural and intermediary determinants of health and health inequity. First, addressing poverty cannot be done in isolation; as complex as it may be, it requires macroeconomic, social, and public policies specifically targeted to vulnerable populations in order to transition them out of poverty. Most critically, cultural and societal values that undermine the role of women and the future of girls may need to be changed to promote gender equity. These structural factors will not only address poverty but also reduce the health inequities experienced by women. As already established, poverty reduction as a Sustainable Development Goal is not possible without the achievement of the other 16 Goals. This makes the achievement of Goal 5, which is aimed at gender equality and the empowerment all women and girls critical for sustainable development. Second, a focus on intermediary social determinants of health, such as the material circumstances of the most vulnerable, behavioral, and biological factors, as well as psychosocial factors, may be important at all levels to ensure gender equality in health. Finally, using a syndemic approach, it is important not only to separately address food, water, or energy insecurity but also to address the economic and political forces that actually tie them together.

References

Aihara, Y. *et al.* (2015) 'Validation of household water insecurity scale in urban Nepal', *Water Policy*, 17(6), pp. 1019–1032. doi: 10.2166/wp.2015.116.

Alaimo, K., Chilton, M. and Jones, S. J. (2020) 'Chapter 17: Food insecurity, hunger, and malnutrition', in Marriott, B. P. *et al.* (eds.) *Present Knowledge in Nutrition* (Eleventh Edition). Cambridge, MA: Academic Press, pp. 311–326. doi: 10.1016/B978-0-12-818460-8.00017-4.

Baiden, P. *et al.* (2020) 'Examining the effects of household food insecurity on school absenteeism among Junior High School students: Findings from the 2012 Ghana global school-based student health survey', *African Geographical Review*, 39(2), pp. 107–119. doi: 10.1080/19376812.2019.1627667.

Barnosky, A. D. *et al.* (2012) 'Approaching a state shift in Earth's biosphere', *Nature*, 486(7401), pp. 52–58. doi: 10.1038/nature11018.

Belachew, T. *et al.* (2011) 'Food insecurity, school absenteeism and educational attainment of adolescents in Jimma Zone Southwest Ethiopia: A longitudinal study', *Nutrition Journal*, 10, p. 29. doi: 10.1186/1475-2891-10-29.

Boateng, G. O. *et al.* (2018) 'A novel household water insecurity scale: Procedures and psychometric analysis among postpartum women in western Kenya', *PLoS One*, 13(6), p. e0198591. doi: 10.1371/journal.pone.0198591.

Boateng, G. O., Balogun, M. R., *et al.* (2020) 'Household energy insecurity: Dimensions and consequences for women, infants and children in low- and middle-income countries', *Social Science & Medicine*, 258, p. 113068. doi: 10.1016/j.socscimed.2020.113068.

Boateng, G. O., Workman, C. L., *et al.* (2020) 'The syndemic effects of food insecurity, water insecurity, and HIV on depressive symptomatology among Kenyan women', *Social Science & Medicine*, p. 113043. doi: 10.1016/j.socscimed.2020.113043.

Boateng, G. O. *et al.* (2021) 'Household energy insecurity and COVID-19 have independent and synergistic health effects on vulnerable populations', *Frontiers in Public Health*, 8. doi: 10.3389/fpubh.2020.609608.

Botreau, H. and Cohen, M. J. (2020) 'Chapter Two: Gender inequality and food insecurity: A dozen years after the food price crisis, rural women still bear the brunt of poverty and hunger', in Cohen, M. J. (ed.) *Advances in Food Security and Sustainability*. pp. 53–117. New York: Elsevier. doi: 10.1016/bs.af2s.2020.09.001.

Brewis, A. *et al.* (2020) 'Household water insecurity is strongly associated with food insecurity: Evidence from 27 sites in low- and middle-income countries', *American Journal of Human Biology*, 32(1), p. e23309. https://doi.org/10.1002/ajhb.23309.

Bruening, M., Dinour, L. M. and Chavez, J. B. R. (2017) 'Food insecurity and emotional health in the USA: A systematic narrative review of longitudinal research', *Public Health Nutrition*, 20(17), pp. 3200–3208. doi: 10.1017/S1368980017002221.

Campbell, B. *et al.* (2017) 'Agriculture production as a major driver of the Earth system exceeding planetary boundaries', *Ecology and Society*, 22(4). doi: 10.5751/ES-09595-220408.

Castillo, D. C. *et al.* (2012) 'Inconsistent access to food and cardiometabolic disease: The effect of food insecurity', *Current Cardiovascular Risk Reports*, 6(3), pp. 245–250. Available at: www.ncbi.nlm.nih.gov/pmc/articles/PMC3357002/ (Accessed: 7 March 2021).

Clean Cooking Alliance (2015) *Women Spend 374 Hours Each Year Collecting Firewood in India, Study Finds, Clean Cooking Alliance*. Available at: www.cleancookingalliance.org/about/news/05-05-2015-women-spend-374-hours-each-year-collecting-firewood-in-india-study-finds.html (Accessed: 7 March 2021).

Collins, S. M. *et al.* (2019) '"I know how stressful it is to lack water!" Exploring the lived experiences of household water insecurity among pregnant and postpartum women in western Kenya', *Global Public Health*, 14(5), pp. 649–662. doi: 10.1080/17441692.2018.1521861.

Cook, J. and Jeng, K. (2009) 'Child Food Insecurity: The Economic Impact on Our Nation', p. 36. Available at https://www.nokidhungry.org/sites/default/files/child-economy-study.pdf (Accessed: 7 March 2021).

Cook, J. T. *et al.* (2008) 'A brief indicator of household energy security: Associations with food security, child health, and child development in US infants and toddlers', *Pediatrics*, 122(4), pp. e867–e875. doi: 10.1542/peds.2008-0286.

Dherani, M. *et al.* (2008) 'Indoor air pollution from unprocessed solid fuel use and pneumonia risk in children aged under five years: A systematic review and meta-analysis', *Bulletin of the World Health Organization*, 86, pp. 390–398C. doi: 10.1590/S0042-96862008000500017.

FAO (1996) *Rome Declaration on World Food Security and World Food Summit Plan of Action*. Rome: FAO Available at: www.fao.org/3/w3613e/w3613e00.htm (Accessed: 3 March 2021).

FAO (2020a) *The State of Food and Agriculture 2020*. Rome: FAO. doi: 10.4060/cb1447en.

FAO (2020b) *The State of Food Security and Nutrition in the World 2020*. Rome: FAO. doi: 10.4060/CA9692EN.

Ferdous, F. *et al.* (2013) 'Severity of diarrhea and malnutrition among under five-year-old children in rural Bangladesh', *The American Journal of Tropical Medicine and Hygiene*, 89(2), pp. 223–228. doi: 10.4269/ajtmh.12-0743.

George, A. S. *et al.* (2020) 'Structural determinants of gender inequality: Why they matter for adolescent girls' sexual and reproductive health', *BMJ*, 368, pp. l6985. doi: 10.1136/bmj.l6985.

Gjerde, J. L. *et al.* (2017) 'Living with pelvic organ prolapse: Voices of women from Amhara region, Ethiopia', *International Urogynecology Journal*, 28(3), pp. 361–366. doi: 10.1007/s00192-016-3077-6.

Gundersen, C. and Ziliak, J. P. (2015) 'Food insecurity and health outcomes', *Health Affairs*, 34(11), pp. 1830–1839. doi: 10.1377/hlthaff.2015.0645.

Hadley, C. and Wutich, A. (2009) 'Experience-based measures of food and water security: Biocultural approaches to grounded measures of insecurity', *Human Organization*, 68(4), pp. 451–460. Available at: www.jstor.org/stable/44148578 (Accessed: 3 March 2021).

Hasegawa, T. *et al.* (2018) 'Risk of increased food insecurity under stringent global climate change mitigation policy', *Nature Climate Change*, 8(8), pp. 699–703. doi: 10.1038/s41558-018-0230-x.

Hernández, D. (2016) 'Understanding "energy insecurity" and why it matters to health', *Social Science & Medicine*, 167, pp. 1–10. doi: 10.1016/j.socscimed.2016.08.029.

International Energy Agency (2017) *Energy Access Outlook 2017: From Poverty to Prosperity*. World Energy Outlook Special Report. Paris: International Energy Agency, pp. 1–140.

Ivers, L. C. and Cullen, K. A. (2011) 'Food insecurity: Special considerations for women', *The American Journal of Clinical Nutrition*, 94(6), pp. 1740S–1744S. doi: 10.3945/ajcn.111.012617.

Jepson, W. *et al.* (2017) 'Progress in household water insecurity metrics: A cross-disciplinary approach', *Wiley Interdisciplinary Reviews: Water*, 4, p. e1214. doi: 10.1002/wat2.1214.

Jones, A. D. (2017) 'Food insecurity and mental health status: A global analysis of 149 countries', *American Journal of Preventive Medicine*, 53(2), pp. 264–273. doi: 10.1016/j.amepre.2017.04.008.

Jung, N. M. *et al.* (2017) 'Gender differences in the prevalence of household food insecurity: A systematic review and meta-analysis', *Public Health Nutrition*, 20(5), pp. 902–916. doi: 10.1017/S1368980016002925.

Kassie, M., Ndiritu, S. W. and Stage, J. (2014) 'What determines gender inequality in household food security in Kenya? Application of exogenous switching treatment regression', *World Development*, 56, pp. 153–171. doi: 10.1016/j.worlddev.2013.10.025.

Krumdieck, N. R. *et al.* (2016) 'Household water insecurity is associated with a range of negative consequences among pregnant Kenyan women of mixed HIV status', *Journal of Water and Health*, 14(6), pp. 1028–1031. doi: 10.2166/wh.2016.079.

Laraia, B. A., Siega-Riz, A. M. and Gundersen, C. (2010) 'Household food insecurity is associated with self-reported pregravid weight status, gestational weight gain, and pregnancy complications', *Journal of the American Dietetic Association*, 110(5), pp. 692–701. doi: 10.1016/j.jada.2010.02.014.

Lee, J. S. and Frongillo, E. A., Jr. (2001) 'Nutritional and health consequences are associated with food insecurity among U.S. elderly persons', *The Journal of Nutrition*, 131(5), pp. 1503–1509. doi: 10.1093/jn/131.5.1503.

Maitra, C. and Food and Agriculture Organization of the United Nations (2018) *A Review of Studies Examining the Link between Food Insecurity and Malnutrition*. Rome: Food and Agriculture Organization of the United Nations. Available at: www.fao.org/3/CA1447EN/ca1447en.pdf (Accessed: 7 March 2021).

Matthew, R. A. (2008) *Resource Scarcity: Responding to the Security Challenge*. New York, NY: International Peace Institute. Available at: www.ipinst.org/wp-content/uploads/2015/06/rscar0408.pdf (Accessed: 3 March 2021).

Miller, J. D. *et al.* (2019) 'Greater household food insecurity is associated with lower breast milk intake among infants in western Kenya', *Maternal & Child Nutrition*, 15(4), p. e12862. https://doi.org/10.1111/mcn.12862.

Milstein, B. (2001) *Introduction to the Syndemics Prevention Network*. Atlanta, GA: Centers for Disease Control and Prevention.

Perkins, J. M. *et al.* (2018) 'Food insecurity, social networks and symptoms of depression among men and women in rural Uganda: A cross-sectional, population-based study', *Public Health Nutrition*, 21(5), pp. 838–848. doi: 10.1017/S1368980017002154.

Popkin, B. M. (2011) 'Contemporary nutritional transition: Determinants of diet and its impact on body composition', *The Proceedings of the Nutrition Society*, 70(1), pp. 82–91. doi: 10.1017/S0029665110003903.

Popkin, B. M., D'Anci, K. E. and Rosenberg, I. H. (2010) 'Water, hydration, and health', *Nutrition Reviews*, 68(8), pp. 439–458. doi: 10.1111/j.1753-4887.2010.00304.x.

Pourmotabbed, A. *et al.* (2020) 'Food insecurity and mental health: A systematic review and meta-analysis', *Public Health Nutrition*, 23(10), pp. 1778–1790. doi: 10.1017/S136898001900435X.

Quinonez, H. M., de Sousa, L. R. M. and Figueroa, L. S. (2019) 'P37 food insecurity and gender inequality in Latin America', *Journal of Nutrition Education and Behavior*, 51(7, Supplement), p. S49. doi: 10.1016/j.jneb.2019.05.413.

Rosinger, A. Y. (2018) 'Household water insecurity after a historic flood: Diarrhea and dehydration in the Bolivian Amazon', *Social Science & Medicine*, 197, pp. 192–202. doi: 10.1016/j.socscimed.2017.12.016.

Sen, A. (1977). Starvation and exchange entitlements: a general approach and its application to the Great Bengal Famine. *Cambridge Journal of Economics* 1(1), 33-59.

Shankar, P., Chung, R. and Frank, D. A. (2017) 'Association of food insecurity with children's behavioral, emotional, and academic outcomes: A systematic review', *Journal of Developmental and Behavioral Pediatrics: JDBP*, 38(2), pp. 135–150. doi: 10.1097/DBP.0000000000000383.

Shariff, Z. M. and Khor, G. L. (2005) 'Obesity and household food insecurity: Evidence from a sample of rural households in Malaysia', *European Journal of Clinical Nutrition*, 59(9), pp. 1049–1058. doi: 10.1038/sj.ejcn.1602210.

Singer, M. and Clair, S. (2003) 'Syndemics and public health: Reconceptualizing disease in bio-social context', *Medical Anthropology Quarterly*, 17(4), pp. 423–441. Available at: www.jstor.org/stable/3655345 (Accessed: 8 March 2021).

Stevenson, E. G. J. *et al.* (2016) 'Community water improvement, household water insecurity, and women's psychological distress: An intervention and control study in Ethiopia', *PLoS One*, 11(4), p. e0153432. doi: 10.1371/journal.pone.0153432.

Tarrass, F. and Benjelloun, M. (2012) 'The effects of water shortages on health and human development', *Perspectives in Public Health*, 132(5), pp. 240–244. doi: 10.1177/1757913910391040.

Tibesigwa, B. and Visser, M. (2016) 'Assessing gender inequality in food security among small-holder farm households in urban and rural South Africa', *World Development*, 88, pp. 33–49. doi: 10.1016/j.worlddev.2016.07.008.

Tsai, A. C. (2018) 'Syndemics: A theory in search of data or data in search of a theory?', *Social Science & Medicine*, 206, pp. 117–122. doi: 10.1016/j.socscimed.2018.03.040.

Tsai, A. C. *et al.* (2016) 'Population-based study of intra-household gender differences in water insecurity: Reliability and validity of a survey instrument for use in rural Uganda', *Journal of Water and Health*, 14(2), pp. 280–292. doi: 10.2166/wh.2015.165.

United Nations Department of Economic and Social Affairs (2015) *The 17 Goals*. Geneva, Switzerland: United Nations. Available at: https://sdgs.un.org/goals (Accessed: 3 March 2021).

UN Millennium Declaration. (2000) A/Res/55/2. New York: United Nations.

US Department of Agriculture, Economic Research Service (2006) *Definitions of Food Security*. Available at: www.ers.usda.gov/topics/food-nutrition-assistance/food-security-in-the-us/definitions-of-food-security.aspx (Accessed: 3 March 2021).

WHO (2014) *Indoor Air Pollution, Health and the Burden of Disease*. Geneva: World Health Organization.

WHO (2019) *Drinking-Water*. Available at: www.who.int/news-room/fact-sheets/detail/drinking-water (Accessed: 7 March 2021).

Widen, E. M. *et al.* (2017) 'Food insecurity, but not HIV-infection status, is associated with adverse changes in body composition during lactation in Ugandan women of mixed HIV status', *The American Journal of Clinical Nutrition*, 105(2), pp. 361–368. doi: 10.3945/ajcn.116.142513.

Widen, E. M. *et al.* (2019) 'HIV infection and increased food insecurity are associated with adverse body composition changes among pregnant and lactating Kenyan women', *European Journal of Clinical Nutrition*, 73(3), pp. 474–482. doi: 10.1038/s41430-018-0285-9.

Wilkinson, P. *et al.* (2009) 'Public health benefits of strategies to reduce greenhouse-gas emissions: Household energy', *The Lancet*, 374(9705), pp. 1917–1929. Available at: www.sciencedirect.com/science/article/pii/S014067360961713X (Accessed: 5 January 2017).

Workman, C. L. and Ureksoy, H. (2017a) 'Water insecurity in a syndemic context: Understanding the psycho-emotional stress of water insecurity in Lesotho, Africa', *Social Science & Medicine*, 179, pp. 52–60. doi: 10.1016/j.socscimed.2017.02.026.

Workman, C. L. and Ureksoy, H. (2017b) 'Water insecurity in a syndemic context: Understanding the psycho-emotional stress of water insecurity in Lesotho, Africa', *Social Science & Medicine*, 179, pp. 52–60. doi: 10.1016/j.socscimed.2017.02.026.

World Data Lab (2021) *World Poverty Clock*. Available at: https://worldpoverty.io (Accessed: 3 March 2021).

Wutich, A. (2009) 'Intrahousehold disparities in women and men's experiences of water insecurity and emotional distress in Urban Bolivia', *Medical Anthropology Quarterly*, 23(4), pp. 436–454. https://doi.org/10.1111/j.1548-1387.2009.01072.x.

Wutich, A. and Brewis, A. (2014) 'Food, water and scarcity: Toward a broader anthropology of resource insecurity', *Current Anthropology*, 55(4), pp. 444–468.

Wutich, A. and Ragsdale, K. (2008) 'Water insecurity and emotional distress: Coping with supply, access, and seasonal variability of water in a Bolivian squatter settlement', *Social Science & Medicine*, 67(12), pp. 2116–2125. doi: 10.1016/j.socscimed.2008.09.042.

9 "Today men's orientation has changed"

Gender and household water and sanitation responsibilities in Ghana

Florence Dery, Meshack Achore, and Elijah Bisung

Introduction

Access to WASH is fundamental to human health and well-being. The lack of clean water and improved sanitation services results in diarrheal diseases, psychosocial distress, and stunted growth and cognitive impairments among children (Dangour et al., 2013; Prüss-Ustün et al., 2014; Spears, Ghosh & Cumming, 2013), posing a severe global health burden. Despite these health challenges, more than 884 million lack access to basic drinking water service globally (WHO & UNICEF, 2017a) and have to resort to alternative (mostly unsafe) sources of water for survival (Achore, Bisung & Kuusaana, 2020). In addition, 2.5 billion people lack access to basic sanitation services (WHO & UNICEF, 2017a). The majority of people without access to basic WASH services live in Low-and-Middle-income-countries (LMICs), particularly SSA (WHO & UNICEF, 2019). In this context, basic water refers to drinking water from a protected source with a collection time of not more than 30 minutes for a round trip, including wait time (WHO & UNICEF, 2017a). On the other hand, basic sanitation refers to the use of improved toilet facilities such as flush/pour flush to piped sewer systems, septic tanks, or pit latrines; ventilated improved pit latrines, compositing toilets or pit latrines with slabs that are not shared with other households (WHO & UNICEF, 2017b). On the other hand, basic hygiene connotes the availability of a handwashing facility on-premises with soap and water (WHO & UNICEF, 2017b). This chapter, however, mostly focuses on basic water and sanitation.

Ghana has significant service gaps in the provision of WASH services, particularly in the area of sanitation. About 13% and 85% of the Ghanaian population lack access to basic water and sanitation, respectively (Ghana Statistical Service, Ghana Health Service, ICF International, 2015), with those in rural areas disproportionately affected (Armah et al., 2018). Over the years, through several policy interventions, including water sector reforms, the government of Ghana has improved overall access to water (GWCL, 2012). However, access to formal water supply in poor urban neighborhoods and rural communities remains a significant challenge. The majority of the urban poor and rural communities without access to regular water supply services turn to unimproved and surface water sources to meet their water needs (Grönwall, 2016; Stoler et al., 2012).

DOI: 10.4324/9780367743918-11

In addition, gender plays a vital role in shaping access to and governance of WASH resources in many low-income countries, including Ghana. Socially constructed gender roles in Ghana have disproportionately placed WASH responsibilities on women and girls. Like most SSA countries, Ghanaian women are primarily responsible for unpaid yet important social reproductive roles, such as collecting water for household use (Ferrant, Pesando & Nowacka., 2014). For instance, a study conducted in 24 countries in SSA by Graham, Hirai, & Kim indicates that adult females are the primary water collectors in many households, and the percentage of households whose primary water collectors are women ranges from 46% (in Liberia) to 90% (in Côte d'Ivoire) (Graham, Hirai & Kim, 2016). The daily search for water also takes a toll on girls' education, as female children were more likely to be responsible for water collection than male children (Graham, Hirai & Kim, 2016). In addition, a recent study by Kangmennaang and colleagues reports that the number of girls within a household level is a strong predictor of water insecurity, as they are primarily responsible for fetching water (Kangmennaang, Bisung & Elliott, 2020). It is common for girls in low-income countries, including Ghana, to wake up before their male siblings fetch water for the household before going to school (Nauges, 2018). Girls are also usually tasked with the responsibility of fetching water after school. Their academic performance understandably suffers as a result (Demie, Bekele & Seyoum, 2016).

Universal and equitable access to WASH, as envisioned in Sustainable Development Goal (SDG) 6, is fundamental to achieving gender equality (SDG 5) in many LMIC contexts, including Ghana (United Nations, 2020). SDG 5 and SDG 6 are inextricably linked due to access to basic drinking water services enabling and accelerating the achievement of gender equality. Thus, the synergistic effects of these SDGs on women and girls' overall health and well-being require greater attention from researchers and WASH practitioners. Drawing on Feminist Political Ecology (FPE) theory, this study critically explores changing perceptions of men's role in household WASH activities and responsibilities in Ghana. Understanding these changing dynamics is necessary for building integrated strategies and policies that create connections between different SDG goals and better understand potential trade-offs among the goals.

Feminist political ecology framework

This research draws on FPE framework. Emerging from Political Ecology (PE), FPE explains how gender is performed and negotiated within environmental struggles and social inequities, including class, caste, race, and religion. FPE highlights the intersections between identity, such as gender, class, ethnicity, and political/economic power relations (Rocheleau, Thomas-Slayter & Wangari, 2013; Buechler & Hanson, 2015). FPE scholars argue that gender predisposes individuals to be environmental and ecological hardships through marginalization and unequal distribution of resources and power (Cole, 2017; Lecoutere, D'Exelle & Van Campenhout, 2015; Resurrección, 2013). The framework calls our attention to gender as an essential factor in human–environmental relationships (Elmhirst, 2011, 2015). It argues that gender intersects with other broader social and political factors to create inequalities. These inequalities result in differing access to WASH services, the individual capacity to adapt to water insecurity, and motivation to engage in WASH activities (Dery et al., 2020).

Feminist scholars from the global south, particularly African feminists, have broadly critiqued the feminist scholarship from the Global North for failing to highlight African women's specific experiences (e.g., oppression) (Adomako Ampofo, Beoku-Betts & Osirim, 2008; Cole, 2017). They argue that African women's access to resources like water, for instance, is negotiated through factors such as patriarchy and power (Cole, 2017). This is important as patriarchy and power are mutually imbued in shaping gender relations (Cole, 2017). The FPE framework is vital in studying WASH services in Ghana and other settings where gender plays a critical role in water management and governance. This study employed FPE in analyzing the interaction between gender and power structures within WASH-related tasks at the household level in Ghana. We draw on the framework to understand the gendered nature of WASH responsibilities, considering the complexity of factors (e.g., social, political, and economic) that contribute to excluding men from adequately participating in household chores.

Furthermore, the FPE framework recognizes that gender does not act in isolation but rather intersects with broader social and political structures to determine if an individual or a group will have access to resources or not (Cole, 2017; Resurrección, 2013). Additionally, FPE theory gives prominence to injustice and inequalities and seeks to locate and explain the causes of inequalities and injustices (Rocheleau, Thomas-Slayter & Wangari, 1996). Finally, the framework pays attention to how gender and other socio-political struggles between men and women or between women intersect to shape their access to and control over resources in a specific context (Sato & Soto Alarcón, 2019).

Study context

The research was conducted in five communities, Gambia, Goamu Kensere, Kwakuri, Ntotroso Wamahinso, and Obengkrom in the Asutifi North District in the Brong Ahafo Region, Ghana. Asutifi North district is one of the 27 districts in the Brong Ahafo Region. The Brong Ahafo Region has since been divided into three regions (Bono, Ahafo, and Oti) after this work was completed. The Asutifi North district is subdivided into four traditional areas (TAs), Kenyasi No. 1, Kenyasi No. 2, Wamanhinso, and Ntotroso. Chiefs administer each TA. Study participants were drawn from all four TAs. The district has a population of 62,816, the majority of which (68%) people reside in rural areas (Asutifi North District, 2014; District Planning Coordinating Unit Kenyasi, 2014)). Subsistence agriculture remains the dominant occupation in the district. Other minor activities include mining and petty trading. About 15% of the district's population lack access to basic water (Asutifi North District Assembly, 2018).

Shared public toilet facilities are commonly used by rural households (50.5%) in the district. However, about 60% of public facilities do not meet the district's threshold of basic sanitation (Asutifi North District Assembly, 2018). Five percent of the district's population practices include open defecation in bushes and fields (Ghana Statistical Service, 2014). In response to the Asutifi North District's WASH inadequacies, a WASH master plan has been developed that sets out key targets and strategies to ensure universal access to basic water and sanitation by 2030 (Asutifi North District Assembly, 2018). The local government, non-governmental organizations, and local authorities from the various TAs support the master plan. Importantly, the study context is a pilot district selected by the International Reference Centre-for WASH (IRC) Ghana (a non-governmental organization) for policy interventions aimed at creating an enabling environment towards achieving the goal of the district master plan.

Methods

Sampling and data collection

Focus group discussions (FGDs) and key informant interviews were used to collect data. We purposefully selected 15 stakeholders (seven men and eight women) as key informants for the study across the five communities (Table 1). The key informants include community leaders, local government professionals, and local WASH practitioners. In addition, ten FGDs were conducted with five women groups and five men groups across the five communities. Each group constituted ten adult respondents except for the community of Ntotroso Wamahinso, where 11 male respondents participated in an FGD, and the community of Obengkrom, which involved 18 FGD participants. The majority of the participants were married with children. All FGD members were from impoverished households without access to basic amenities, including safe water and toilet facilities at home.

Both FGDs and key informant interviews were audio-recorded with participants' consent. To encourage each participant's full understanding and participation, they were interviewed exclusively in the language in which they were most comfortable. Each key informant interview lasted approximately between 30 and 60 minutes. Information was collected concerning norms and practices around household chores, including WASH-related tasks (e.g., water collection, cleaning, sweeping) and challenges in accessing water and sanitation services. Key informant interviews took place in various locations, including participants' workplaces, public places, and homes. The FGDs, on the other hand, which lasted 45–90 minutes, occurred in public places and were facilitated by two researchers. Discussions focused on prevailing norms and practices, household members' rights and responsibilities regarding household chores, and water and sanitation access challenges. The study received ethics clearance from the Queen's University's Health Sciences and Affiliated Teaching Hospitals Research Ethics Board, who approved the study protocol, including the informed consent process.

Data analysis

The study adopted an inductive approach to qualitative data analysis, and specifically thematic analysis (Guest, MacQueen & Namey, 2014). The following steps were followed in analyzing the data: 1) digital recordings of the key informant interviews and FGDs were transcribed in the same language in which the interview was conducted, de-identified, and then translated into English if they were not conducted in English; 2) data were carefully examined, and significant statements (sentences/quotes) emerging from the data were extracted to demonstrate patterns and elements of WASH responsibilities; 3) related statements were categorized and coded under broad themes, and finally; 4) themes were then used to write the differential perceptions of men and women as they pertain to WASH responsibilities. These processes were completed in NVivo 15, a qualitative analysis software.

Results

In this section, we report our analysis results regarding factors at the community and household levels that influence how people perceive WASH responsibilities. Our study showed that poverty, distribution of unpaid labor, and gender-based power dynamics strongly affect individuals' perception of WASH responsibilities. We use pseudonyms in reporting our results to ensure participants' anonymity.

Poverty

Our findings revealed that poor households were more likely to push WASH responsibilities on women. A female participant, for instance, recounted:

> Some men will tell you they do not have adequate money, and so you the woman, will have to contribute money to support the construction of a toilet or bathroom at home; if not it will not work. Everyone knows in this community that does all kinds of work, including what men do. I take all the decisions at home regarding the construction of toilets.
>
> (Talata, FGD)

Another participant reported:

> We are suffering in this community due to low income and unemployment. The money for housekeeping is provided by our husbands, including money we pay for fetching water. It is the responsibility of us women to manage the money and water well.
>
> (Adjoa, FGD)

Although participants from low households also desired water and toilet facilities, they mentioned that they were struggling to fulfill their basic needs. Hence, their immediate priority was not the construction of a water source or toilet. One participant indicated how inadequate financial resources impact acceptable hygiene practices:

> The hardship in this community is too much. One is aware of the excellent WASH practices but cannot afford them. We do not even have adequate money to buy food; how much more soap.
>
> (Kwesi, key informant)

Furthermore, other participants reiterated the impact of lack of access to water and toilet facilities because of lack of financial resources:

> In this big community, we have only about six public toilet facilities, which is not enough to meet our demands and needs. These facilities include WCs (water closets) and the irregular flow of water, making them not accessible most of the time. We also have the VIPs (Kumasi ventilated improved pits). Due to the inadequacy of the public toilet facilities, some people have to resort to unapproved ways of answering nature's call. This is not dignifying as it results in many diseases. Only about 10% can afford to put up private toilet facilities in their homes.
>
> (Akweley, key informant)

Another stated how inadequate financial resources affect access to safe water:

> We use pipe water for drinking, cook, and bath. Where I live, we buy the water for 30 pesewas per yellow gallon from a privately owned pipe. I fetch water from the unprotected well to take care of my animals. The pipe water does not flow regularly and lacks the money to buy the water . . . so sometimes it is difficult to get water, and one needs to cut down on how much water one uses to survive. For me, sometimes I fetch the water with permission to pay later, but not everyone is given such permission.
>
> (Kekeli, key informant)

Many women also reported that easy access to water and sanitation facilities could reduce their daily struggles. Particularly, married women with children mentioned that toilets and bathroom facilities at home could reduce their burden of always collecting water and cleaning gathering feces littered around the house:

> Some of us get tired before we go to the farm. We get up and walk long distances to collect water. This makes us tired before we go to the farm. If we could get water connected to us at home, we will not spend lots of time collecting water and accessing toilet facilities if we get a toilet at home. We will not get too tired, and our health will improve.
>
> (Ama, FGD)

The study revealed that toilet or bathroom construction might not be a priority in some households, even if they have adequate resources to construct them due to men not perceiving them as a priority. A woman complained:

> If you, the woman, have little kids at home who cannot access the toilet facility far away from home, you know how helpful a toilet facility at home could be if the man does not know. Therefore, the woman can tell the man how unsanitary it is when the children defecate in the open around the house and contaminate everything at home that could result in several diseases. . . . So

> if these women do not suggest the construction of a bathroom, the men do not mind.
>
> (Lizzy, FGD)

Access to adequate water and sanitation has improved over time in the study area, but pockets of deprivation remain. Besides geographical disparities, lack of access to adequate water and improved sanitation is mostly associated with poverty.

Unequal distribution of unpaid labor

This theme captured household, cultural, and socio-economic structures that influenced the unequal distribution of unpaid labor. Household composition and structure are likely to alter the family's decision-making power and distribution of domestic responsibilities. As reported by one woman in a focus group:

> I am a widow taking care of 6 children. I make all decisions including the children's feeding and school fees. I do not have any support, so I work tirelessly. I am not even 40 years old, but I look like an old woman, all because of too much stress.
>
> (Nana Ama, FGD)

Another participant indicated:

> Lately, not only girls perform household chores. In my case I gave birth to 8 children, but only one is a girl. I had the boys first before the girl, and so the boys do the chores, as one cooks, one cleans, and the others collect water. They all sweep every morning.
>
> (Serwa, FGD)

Most women in all the discussion groups indicated that meeting their WASH needs is largely dependent on their efforts, as men in many cases do not prioritize labor in these areas. As stated by one of the female focus group participants:

> The men do not care if there is no bathroom or toilet facility at home. Most women in this community construct bathrooms at home. We go to the forest and carry wood. We also sell farm produce in order to buy nails. With the wood and nails, we are able to put up bathrooms for the household as it is important for us.
>
> (Akweley, FGD)

Shifts regarding social and economic forces, such as: shifts in educational opportunities and achievement, paid employment in formal and informal labor markets, and migration, affect male perceptions regarding the distribution of responsibilities at home. One of the focus group participants stated that:

> In my opinion, I see more empowered men today than before. Due to modernization and education, there are more empowered men than it was in the

> olden days. For example, in the olden day, men believed that women are responsible for all household chores, including fetching water. However, today men's orientation has changed, and they are seen doing household chores that were not previously acceptable.
>
> (Boakye, FGD)

Family farming is common in the study area and is mostly characterized by mixed farming. Men mostly hold greater decision-making power regarding farming activities. In addition to being responsible for household chores, including collecting water and taking care of children, the women offer a helping hand on the farm. The importance of women's contribution to the family farm's success cannot be ignored and hence motivates male participation in household chores. Because of the role women play in the family farms, husbands are motivated to help women with their chores (which are mostly WSH-related activities) to lend them (men) a helping hand in the farms. For instance, a male participant stated:

> We help the women that were previously frowned upon because we assist them in sweeping and collecting water to finish on time and go to the farm together. It is nice to walk with your wife to the farm.
>
> (Fifi, FGD)

Some participants also feel that co-responsibility for household chores is more acceptable in urban settings than rural areas. According to one of the FGD participants:

> As for sharing responsibilities between men and women, it is more prominent in the big cities. In the cities, men collect water and assist their wives.
>
> (Atta, FGD)

In addition, most women and men reported customary or cultural norms as barriers to division of labor at the household and community levels. One key informant describes this as follows:

> Mostly, tradition demands that women are solely responsible for WASH activities. . . . Women have every right to suggest to their men to build toilet facilities because it is difficult for women to engage in open defecation. However, the man makes the ultimate decision of having the toilet built.
>
> (Kwesi, key informant)

Though traditional values dominate, men are more likely than ever to pick up some WASH-related chores at the household level.

Gender-based power dynamics

It is mostly assumed that households make decisions about WASH activities as a single unit. However, the reality is that men and women have different abilities when it comes to making decisions at the household level. Intra-household power discrepancies exist and mostly follow ascribed gender norms and gender roles. These discrepancies influence decision-making processes, with women having limited roles in decision-making. All participants agreed that there are varied reasons for constructing a water source, toilet, and bathroom at home amidst poverty. Many women indicated that the decision to construct a WASH facility resided with their male spouses, who would only construct them if they felt it was necessary, implying women had a limited role in decision-making at the household level. A male participant stated how important access to WASH is a critical concern for women safety:

> I can also defecate openly without fear. However, the women cannot. Therefore, when the women say we need to put up a bathroom, then the man and the woman can deliberate on the way forward. Even with regard to sitting on the toilet or bathroom, women's opinions should be highly considered.
>
> (Gyan, FGD)

However, participants indicated that WASH services could contribute to a positive change in gender relations. Women in community WASH committees were found to have increased agency. This mostly originates from the specific WASH empowerment training offered to women and men on human rights and gender equality by state and non-governmental organizations. Women were given a platform to share their ideas, which significantly contributed to women having more confidence to negotiate roles and make decision-making in their homes. A woman explained:

> Years ago, my husband used to make decisions on everything in our home. Now, he listened to me and built the toilet because I told him that it prevents people from seeing my nakedness.
>
> (Eve, key informant)

A male participant also stated:

> Through the WASH training program organized in the community, my wife is more knowledgeable in maintaining a clean environment and caring for everyone. Therefore, if a woman like my wife comes up with the suggestion of investing in a toilet facility, there is no good man who would ignore such a suggestion.
>
> (Mensa, FGD)

A female participant also stated:

> A section of the community was to be resettled in a different location. Individuals from the affected group were asked to form a committee to see the development of housing facilities in the new settlement area. Women were not included in the formation of the committee. After the completion of the

> housing facilities, I was part of the assessment group to assess the state of the completed houses. We realized that the houses were well constructed but without a kitchen and toilets. If women were part of the resettlement committee, these serious omissions would have been avoided. In the subsequent formation of resettlement committees, women were included, and as such, houses built later on had toilets and kitchens.
>
> (Shine, FGD)

Though the study area is predominantly a farming community, some mining activities are being done. Part of the community is resettled to pave the way for mining activities.

In addition, improvements in WASH infrastructure have changed household work's nature, leading to a reduction in the burden of domestic chores women are challenged to do while saving time for other relevant activities, such as economic activities. Thus, some men are beginning to take on some of this work, thereby redistributing household roles. One male participant reported:

> I am taking myself as an example. I live with only my wife. I help with the chores; while she is sweeping, I fetch water from the compound. Doing this helps us go to the farm early.
>
> (Oduro, FGD)

Some participants without access to WASH infrastructure at home also reported a positive change in gender relations. A female key informant stated:

> Women are mostly responsible for carrying out WASH activities such as fetching water, cleaning. Men are willing to help. In fact, some men help their wives. For instance, my husband sometimes fetches water from a nearby village using his motorbike when faced with severe water shortage at home.
>
> (Naana, key informant)

However, it is also important to note that improvements in WASH infrastructure alone would not necessarily change power dynamics between women and men if unfavorable social norms remain. It may even exacerbate the burden women face in carrying out household chores. One male participant reported:

> A good woman will ensure that her responsibilities are carried out well at all times. Therefore, I will not condone laziness if she cannot fulfill her responsibility of getting enough water as expected of every woman in this society. I will therefore use water as I wish, irrespective of anything. It is her sole duty to do all domestic chores, including collecting water for us to use.
>
> (Opanyin-Kojo, FGD)

Another male participant reported:

> Mostly, tradition demands that women are solely responsible for WASH activities. Even if a man is living alone, he finds a woman to come and clean his bathroom for him. Per our traditions, men are not supposed to clean the bathroom.
>
> (Kwesi, key informant)

Another male participant reported how doing WASH-related chores affect one's masculinity. He stated:

> If people see a man sweeping early in the morning, they lose respect for you and see you to be less of a man. Because of this, men shy away from participating in such chores just to avoid being ridiculed in society.
>
> (Ansah, FGD)

The involvement of traditional leaders in WASH activities was also reported as a contributing factor to changes in gender relations at home. Such people in authority can encourage equal participation of women and men in WASH activities, as well as increase the awareness of the importance of gender equality at the household level. One woman stated:

> Women and men engage themselves in the WASH activities. In some cases, we have our queen mothers coming out to lead.
>
> (Susan, key informant)

Increasing investments in WASH services facilitate women's participation in decision-making at the household level and reduce unpaid work on women and girls.

Discussion

This study investigates the gendered nature of WASH responsibilities and the changing perceptions of men's role in improving WASH services at the household and community level. Out thematic qualitative data suggest that poverty, distribution of unpaid labor, and gender-based power dynamics at the household and community levels significantly influence how people perceive WASH responsibilities. Our findings show gender differences in terms of WASH responsibilities, particularly in poor rural households in Ghana without WASH facilities. WASH responsibilities are shouldered by women and girls who, in most cases, walk long distances to collect water. These findings align with those of previous studies (Graham, Hirai & Kim, 2016; Pouramin Nagabhatla & Miletto, 2020). For instance, Graham, Hirai and Kim (2016) illustrated that water collection practices are gendered, with women shouldering more responsibilities. Although some men are beginning to assist women with water collection at the household level, they do so with motorbikes and bicycles' assistance. However, most women do not have access to these resources, which makes water transportation a more tedious task for women. This unequal access to transportation resources results in women spending more of their otherwise productive time on water collection. In addition to the economic impact of shouldering household water responsibilities, walking for long distances to fetch water reportedly has biophysical effects, such as bodily pains (Venkataramanan et al., 2020), musculoskeletal trauma (Pouramin Nagabhatla & Miletto, 2020), and psychosocial stress from daily struggles and efforts to meet WASH needs (Brewis, Choudhary & Wutich, 2019; Stevenson et al., 2016). Thus, women are disproportionately at higher risk of being exposed to waterborne diseases and musculoskeletal trauma resulting from the lack of easy access to water services (Geere et al., 2018).

Our findings also indicate that although primarily responsible for WASH activities, women are excluded from partaking in WASH-related decisions. Men are mostly considered the providers and head of households in Ghana and are the primary decision makers. Decisions made by women have to be approved by the head of the household (men). As illustrated, many women cannot swiftly take decisions to construct toilet facilities at home; they need approval from their husbands and, in some instances, fathers. This is due to women's lack of financial freedom. Many of the female FGD participants strongly indicated their interest in having the independent right to make WASH-related decisions, given that they are mostly affected by the outcomes of these decisions. However, most are unable to exercise these rights due to financial and socio-cultural restrictions.

Economic and cultural norms also restrict women's participation in community-level activities. For instance, women are seen as lacking the technical skills to hold WASH-related managerial positions or partake in WASH committees. This consequently widens the inequality gap between men and women. To achieve SDG 5, it is imperative to accord women the opportunity to participate in household decisions, particularly WASH-related decisions (Kilsby, 2012). Apart from sensitizing communities on gender equality, empowerment training allows women to realize that their agency will boost women's participation in decision-making and community WASH-related activities. In doing so, WASH practitioners should recognize the crucial role that men play in bringing about positive changes in gender relations and involve them in their activities. Doing so will enable both men and women to work together to achieve WASH goals at the household and community levels. Excluding men in women empowerment activities results in unintended effects, such as gender-based violence, as men might misconstrue some of the activities or decisions as challenging their masculinity (Gupta et al., 2013).

The findings reveal that changing perceptions and socio-cultural norms may increase men's participation in household WASH activities. Leveraging community leaders' influence to help change social and cultural norms that prevent the uptake of household WASH responsibilities should not be overlooked. Traditional leaders and authorities play an essential role and can also be engaged in changing perceptions of WASH-related behaviors. This can create an enabling environment for men to challenge traditional norms by being able to engage in household activities that were once reserved for women. Together with local authorities, traditional leadership could serve as powerful tools to educate people about the significance of shared WASH responsibilities.

Limitations

This study had a number of worth acknowledging limitations, particularly in the interpretation of the findings. First, selection bias is possible, as our sample was limited to participants from organizations and community groups within the study district. Second, the translation process of the interviews has the potential to affect the meaning of local expressions. Lastly, the study focused on collecting qualitative information about a specific setting with certain demographic characteristics. While this provided insightful information, this may not be generalizable to contexts with significantly different demographics.

Conclusions

This chapter highlights gender inequalities in WASH-related household activities by drawing our attention to the intersection between gender and other factors such as class, culture, and power relations. Thus, in implementing policy interventions, practitioners should take into consideration how broader socio-cultural and political factors create inequalities and how these inequalities result in differing access to WASH between men and women. As practitioners look for solutions to WASH problems, the needs of the people (women) who shoulder WASH responsibilities should take precedence by listening to and incorporating their voices in WASH-related activities and decisions at both the community and household levels. Communities and local institutions, particularly those that work in the WASH sector, should address social and cultural barriers that push WASH activities' burden at the household level on women. This can be achieved by creating training and empowerment programs to address discrimination against women and promote gender equality.

The SDG 5 and SDG 6 are inextricably linked, and their synergistic effects on the overall health and well-being of women and girls require greater attention from researchers and WASH practitioners. Inasmuch as it is important to make safe drinking water available to all, achieving this will only be possible if men and women actively partake in WASH-related activities at the household and community levels. That is, gender equality, as in SDG 5, is a prerequisite to the overall success of the SDG 6 agenda; investing in gender equality where there is equal participation of men and women can deliver high returns in terms of universal access to safe drinking water.

References

Achore, M., Bisung, E., & Kuusaana, E. D. (2020) Coping with water insecurity at the household level: A synthesis of qualitative evidence. *International Journal of Hygiene and Environmental Health*, 230, 113598. https://doi.org/10.1016/j.ijheh.2020.113598

Adomako Ampofo, A., Beoku-Betts, J., & Osirim, M. J. (2008) Researching African women and gender studies: New social science perspectives. *African and Asian Studies*, 7(4), 327–341. https://doi.org/10.1163/156921008X359560

Armah, F. A., Ekumah, B., Yawson, D. O., Odoi, J. O., Afitiri, A. R., & Nyieku, F. E. (2018) Access to improved water and sanitation in sub-Saharan Africa in a quarter century. *Heliyon*, 4(11), e00931. https://doi.org/10.1016/j.heliyon.2018.e00931

Asutifi North District. (2014) *District Medium-Term Development Plan*. Available from: https://new-ndpc-static1.s3.amazonaws.com/CACHES/PUBLICATIONS/2016/04/04/ER_Asutifi+North_2014-2017+DMTDP.pdf (Accessed: 20 December 2020).

Asutifi North District Assembly 2018 *Water Sanitation and Hygiene (WASH) Masterplan*. Asutifi North District Assembly, Kenyasi, Ghana. Available from: www.ircwash.org/resources/water-sanitation-andhygiene-wash-masterplan-asutifi-north-district-ghana (Accessed: 10 September 2020).

Brewis, A., Choudhary, N., & Wutich, A. (2019) Household water insecurity may influence common mental disorders directly and indirectly through multiple pathways: Evidence from Haiti. *Social Science & Medicine*, 238, 112520.

Buechler, S., & Hanson, A. M. (2015) A political ecology of women, water and global environmental change. In *A Political Ecology of Women, Water and Global Environmental Change*. London, UK: Routledge. https://doi.org/10.4324/9781315796208

Cole, S. (2017) Water worries: An intersectional feminist political ecology of tourism and water in Labuan Bajo, Indonesia. *Annals of Tourism Research*, 67, 14–24. https://doi.org/10.1016/j.annals.2017.07.018

Dangour, A. D., Watson, L., Cumming, O., Boisson, S., Che, Y., Velleman, Y., Cavill, S., Allen, E., & Uauy, R. (2013) Interventions to improve water quality and supply, sanitation and hygiene practices, and their effects on the nutritional status of children. *Cochrane Database of Systematic Reviews*, (8). https://doi.org/10.1002/14651858.CD009382.pub2

Demie, G., Bekele, M., & Seyoum, B. (2016) Water accessibility impact on girl and women's participation in education and other development activities: The case of Wuchale and Jidda Woreda, Ethiopia. *Environmental Systems Research*, 5(1), 1–12. https://doi.org/10.1186/s40068-016-0061-6

Dery, F., Bisung, E., Dickin, S., & Dyer, M. (2020) Understanding empowerment in water, sanitation, and hygiene (WASH): A scoping review. *Journal of Water, Sanitation and Hygiene for Development*, 10(1), 5–15. https://doi.org/10.2166/washdev.2019.077

District Planning Coordinating Unit Kenyasi. (2014) *District Medium-Term Development Plan*. Available from: https://new-ndpc-static1.s3.amazonaws.com/CACHES/PUBLICATIONS/2016/04/04/ER_Asutifi+North_2014-2017+DMTDP.pdf (Accessed: 12 August 2020).

Elmhirst, R. (2011) Introducing new feminist political ecologies. *Geoforum*, 42(2), 129–132. https://doi.org/10.1016/j.geoforum.2011.01.006

Elmhirst, R. (2015) Feminist political ecology. In *The Routledge Handbook of Gender and Development*. London, UK: Routledge. https://doi.org/10.4324/9780203383117

Ferrant, G., Pesando, L. M., & Nowacka, K. (2014) Unpaid care work: The missing link in the analysis of gender gaps in labour outcomes. *OECD Development Centre*.

Geere, J. A., Bartram, J., Bates, L., Danquah, L., Evans, B., Fisher, M. B., Groce, N., Majuru, B., Mokoena, M. M., Mukhola, M. S., Nguyen-Viet, H., Duc, P. P., Williams, A. R., Schmidt, W. P., & Hunter, P. R. (2018) Carrying water may be a major contributor to disability from musculoskeletal disorders in low-income countries: A cross-sectional survey in South Africa, Ghana and Vietnam. *Journal of Global Health*, 8(1). https://doi.org/10.7189/jogh.08.010406

Ghana Statistical Service. (2014) *2010 Population & Housing Census: District Analytical Report: Asutifi North District.*

Ghana Statistical Service, Ghana Health Service, ICF International. (2015) *Ghana Demographic and Health Survey 2014*. Accra: Ghana Statistical Service.

Graham, J. P., Hirai, M., & Kim, S. S. (2016) An analysis of water collection labor among women and children in 24 sub-Saharan African countries. *PLoS One*, 11(6), e0155981. https://doi.org/10.1371/journal.pone.0155981

Grönwall, J. (2016) Self-supply and accountability: To govern or not to govern groundwater for the (peri-) urban poor in Accra, Ghana. *Environmental Earth Sciences*, 75(16), 1–10. https://doi.org/10.1007/s12665-016-5978-6

Guest, G., MacQueen, K. M., & Namey, E. E. (2014) *Applied Thematic Analysis*. Thousand Oaks, CA: Sage Publications.

Gupta, J., Falb, K. L., Lehmann, H., Kpebo, D., Xuan, Z., Hossain, M., Zimmerman, C., Watts, C., & Annan, J. (2013) Gender norms and economic empowerment intervention to reduce intimate partner violence against women in rural Côte d'Ivoire: A randomized controlled pilot study. *BMC International Health and Human Rights*, 13(1), 1–12. https://doi.org/10.1186/1472-698X-13-46

Ghana Water Company (GWCL). 2012. *History of Water Supply in Ghana*. Accra, Ghana: GWCL. Available: http://www.gwcl.com.gh/pgs/history.php (Accessed : 26 June 2020).

Kangmennaang, J., Bisung, E., & Elliott, S. J. (2020) 'We are drinking diseases': Perception of water insecurity and emotional distress in urban slums in Accra, Ghana. *International Journal of Environmental Research and Public Health*, 17(3), 890. https://doi.org/10.3390/ijerph17030890

Kilsby, D. (2012). Now We Feel Like Respected Adults': Positive Change in Gender Roles and Relations in a Timor-Leste WASH Program, research conducted by International Women's Development Agency. Melbourne, WaterAid Australia, Melbourne and WaterAid Timor-Leste, Dili. https://doi.org/10.1007/s00520-008-0463-8

Lecoutere, E., D'Exelle, B., & Van Campenhout, B. (2015) Sharing common resources in patriarchal and status-based societies: Evidence from Tanzania. *Feminist Economics*, 21(3), 142–167. https://doi.org/10.1080/13545701.2015.1024274

Nauges, C. (2018) Correction to: Water hauling and girls' school attendance: Some new evidence from Ghana. *Environmental and Resource Economics*, (2017), 66(1), 65–88. https://doi.org/10.1007/s10640-015-9938-5. In *Environmental and Resource Economics*. https://doi.org/10.1007/s10640-017-0194-8

Pouramin, P., Nagabhatla, N., & Miletto, M. (2020) A systematic review of water and gender interlinkages: Assessing the intersection with health. *Frontiers in Water*, 2, 6. https://doi.org/10.3389/frwa.2020.00006

Prüss-Ustün, A., Bartram, J., Clasen, T., Colford, J. M., Cumming, O., Curtis, V., Bonjour, S., Dangour, A. D., De France, J., Fewtrell, L., Freeman, M. C., Gordon, B., Hunter, P. R., Johnston, R. B., Mathers, C., Mäusezahl, D., Medlicott, K., Neira, M., Stocks, M., . . ., & Cairncross, S. (2014) Burden of disease from inadequate water, sanitation and hygiene in low- and middle-income settings: A retrospective analysis of data from 145 countries. *Tropical Medicine & International Health*, 19(8), 894–905. https://doi.org/10.1111/tmi.12329

Resurrección, B. P. (2013) Persistent women and environment linkages in climate change and sustainable development agendas. In *Women's Studies International Forum* (Vol. 40, pp. 33–43). Pergamon. https://doi.org/10.1016/j.wsif.2013.03.011

Rocheleau, D., Thomas-Slayter, B., & Wangari, E. (1996) Gender and environment: A feminist political ecology perspective. In *Feminist Political Ecology: Global Issues and Local Experience*. London: Routledge. (pp. 3–26)

Rocheleau, D., Thomas-Slayter, B., & Wangari, E. (Eds.). (2013) *Feminist Political Ecology: Global Issues and Local Experience*. Abingdon, UK: Routledge.

Sato, C., & Soto Alarcón, J. M. (2019) Toward a postcapitalist feminist political ecology' approach to the commons and commoning. *International Journal of the Commons*, 13(1). https://doi.org/10.18352/ijc.933

Spears, D., Ghosh, A., & Cumming, O. (2013) Open defecation and childhood stunting in India: An ecological analysis of new data from 112 districts. *PLoS One*, 8(9), e73784. https://doi.org/10.1371/journal.pone.0073784

Stevenson, E. G. J., Ambelu, A., Caruso, B. A., Tesfaye, Y., & Freeman, M. C. (2016) Community water improvement, household water insecurity, and women's psychological distress: An intervention and control study in Ethiopia. *PLoS One*, 11(4), e0153432.

Stoler, J., Fink, G., Weeks, J. R., Otoo, R. A., Ampofo, J. A., & Hill, A. G. (2012) When urban taps run dry: Sachet water consumption and health effects in low-income neighborhoods of Accra, Ghana. *Health & Place*, 18(2), 250–262. https://doi.org/10.1016/j.healthplace.2011.09.020

United Nations. (2020) *The 17 Goals Sustainable Development*. New York, NY: United Nations. Department of Economic and Social Affairs.

Venkataramanan, V., Geere, J. A. L., Thomae, B., Stoler, J., Hunter, P. R., & Young, S. L. (2020) In pursuit of 'safe' water: The burden of personal injury from water fetching in 21 low-income and middle-income countries. *BMJ Global Health*, 5(10), e003328.

WHO, & UNICEF. (2017a) Progress on drinking water, sanitation and hygiene: 2017 update and SDG baseline. *World Health Organization*. https://doi.org/10.1016/j.pnpbp.2017.06.016. (Accessed on 10 June 2020).

WHO, & UNICEF. (2017b) Progress on drinking water, sanitation and hygiene: Launch version July 12 main report progress on drinking water, sanitation and hygiene. *WHO Library Cataloguing in Publication Data*. Available from: https://www.who.int/mediacentre/news/releases/2017/launch-version-report-jmp-water-sanitation-hygiene.pdf. (Accessed on 10 June 2020).

WHO, & UNICEF. (2019) Progress on household drinking water, sanitation and hygiene 2000–2017 special focus on inequalities. *Launch Version July 12 Main Report Progress on Drinking Water, Sanitation and Hygiene*.Available from: https://www.unicef.org/reports/progress-on-drinking-water-sanitation-and-hygiene-2019. (Accessed on 10 June 2020).

10 Canvas Totes and plastic bags

The political ecology of food assistance effectiveness at farmers' markets in the Twin Cities

Sophia Alhadeff and William G. Moseley

Introduction

[1] At the end of June 2019, the Trump Administration proposed a policy that could result in three million people losing access to food stamp benefits (Fessler, 2019). The Supplemental Nutrition Assistance Program (SNAP), formerly known as the food stamp program, is a federal food aid program designed to help low-income individuals and families combat food insecurity in the United States. In the Twin Cities, Minnesota, USA, SNAP benefits are accepted at a selection of farmers' markets to expand the accessibility of fresh, local produce and handmade products.

This chapter explores the effectiveness of SNAP usage at farmers' markets in the Twin Cities and the ways in which it can be improved to result in better access to healthy, nutritious, and culturally appropriate food choices. While research in this field has examined various coping strategies used to combat food insecurity (Larson and Moseley, 2012), this project focused on the role of farmers' markets as a means of providing local, healthy and affordable food as well as the role of gender in farmers market spaces (HLPE, 2020). This framing places the study squarely in the context of two of the UN sustainable development goals, #2 on ending hunger and improving nutrition and #5 on advancing gender equality and empowering women. In an environment where food justice movements are often steeped in elitism, this project aims to provide ideas to democratize farmers' markets to better serve low-income communities who are often marginalized within the food system (Slocum, 2007; Gottlieb and Joshi, 2010). In this chapter, we explore the question: How effective is the use of government food assistance programs at farmers' markets for the purpose of combating food and nutrition insecurity[2] in the Twin Cities? Using a feminist political ecology framework, we examine three farmers' markets across the Twin Cities to evaluate the effectiveness of SNAP in this context. By conducting semi-structured interviews from shoppers, market managers, and key informants, we aim to better understand how people interact with the farmers' market setting. We argue that both physical and social accessibility influences the effectiveness of SNAP use at farmers' markets in the Twin Cities and thus propose two policy recommendations to improve both factors.

DOI: 10.4324/9780367743918-12

This chapter first contextualizes the study within the scholarly literature, before providing relevant background on the history and development of food assistance policy in the United States and the state of Minnesota. The methods of our study are then outlined and then the results of three markets are discussed. The findings are analyzed in relation to social accessibility, to the trends in the literature, as well as SDGs #2 and #5. Policy recommendations with steps for implementation on multiple scales are offered.

Context in the literature

In order to analyze the results of our study and answer our guiding research questions, we employ feminist political ecology and embodiment geography frameworks. Feminist political ecology highlights social difference within a particular situation (Rocheleau, 1996). Moreover, feminist political ecology acknowledges gender power dynamics as well as the distinction between socially constructed gender roles and physical biology while also moving away from a gender binary toward a continuum (Rocheleau, 1996). Feminist political ecology is useful in analyzing food assistance and famers' market literature as they recognize structure, marginalization, and social difference, such as race, class, gender, sexuality, and able-bodiedness, as key aspects to improving food assistance at farmers' markets. Embodiment geography offers "a discussion of the materiality of race [or other identities] rather than its representation or performance" (Slocum, 2008, p. 849). This growing field of scholarship explores sensual, emotional, or affective geographies and lends itself to understand how individuals interact with their environment based on their identity (Clifford et al., 2010). Together, feminist political ecology and embodiment geography provide a framework to best analyze the current academic conversion around food access at farmers' markets and effective food policy, including discussions about how best to address SDG #2 on ending hunger and SDG#5 on advancing gender equality.

A recent UN report highlights the importance of local food production, farmer's markets and greater attention to gender power dynamics as critical steps to address SDG#2 to end hunger and reduce malnutrition in all its forms by 2030 (HLPE, 2020). Critics of the SDG process argue that we have failed to make progress on SDG #2 to date because we have been too focused on increasing food production and have not paid sufficient attention to improving access to healthy and nutritious food (Moseley, 2018). Furthermore, while we sometimes conceive the SDGs as discrete targets, they are often interlinked. As such, to the extent that women play pivotal roles in food systems as producers, vendors, shoppers and cooks, then empowering women (or making advances on SDG #5) often facilitates progress on addressing hunger and malnutrition (SDG #2) (ILO, 2017; Rao, 2020). There are eight targets related to achieving SDG#2, for which two are directly relevant to this study. The first target highlights the importance of access (including farmers' markets) to addressing hunger and malnutrition, and the third target highlights the role of women in food systems (UN, 2016). There are nine targets related to SDG #5, for which the fourth target highlights the importance of recognizing and valuing unpaid domestic work by women, and providing public services, infrastructure, and social protection policies to support this work (UN, 2016). Since women play a central role in feeding their families (and much of this labor is unpaid), this SDG target clearly highlights the importance of social protection programs like SNAP and farmers' market infrastructure (Shannon, 2014, 2016). Below we identify gaps in the literature and where improvements can be made to US food assistance policy.

Why farmers' markets?

While there is some speculation as to if farmers' markets are adequate methods of food justice, the general conclusion in the academic literature is that they function as a successful method of connecting shoppers to local producers (Gottlieb and Joshi, 2010; Winne, 2008). Additionally, farmers' markets are fully incorporated into the discussion of the food justice and local food movements. Farmers' markets have often been criticized for perpetuating inequality within the food system. With the rise of popularity, farmers' markets gained a reputation for higher cost items (Gottlieb and Joshi, 2010, p. 166).

A barrier to farmers' markets exists based on the assumption that the environment is designed for the desires of upper-middle-class, white shoppers who want to engage in the local food system as an activity rather than as a regular means of acquiring food. According to Julie Guthman, geographer and food scholar, "it is in inequalities that neoliberalism has exacerbated that reveals the limits of the alternative food movement" (Guthman, 2011, p. 164). It is also crucial to recognize the vulnerability of farmers' markets. As farmers' markets have increased in popularity, resources have been spread thin in most neighborhoods (Stephenson et al., 2008, p. 188).

Social barriers to accessing farmers' markets

There are a multitude of social factors that also impede access. By employing a feminist political ecology lens, scholars evaluate factors of social difference and identity in the farmers' market landscape. In a Twin Cities-based study, Rachel Slocum evaluates racialized food practices through the implementation of two themes of embodiment, racial division and intimacy, at the Minneapolis Farmers Market. Slocum employs Elizabeth Grosz's corporeal feminist theory, which refers to "a dynamic capacity of human bodies to emerge in relation to each other" (Slocum, 2008, p. 853). With the acknowledgment that race is an embodied practice, Slocum investigates divisions at the markets, ways of interacting in the market space, food preferences, and other behavioral patterns that fall along racial lines (Slocum, 2008). Moreover, Slocum discusses how bodies present themselves distinctly and how market tourism is an emerging trend among white, middle-class groups that alienates communities of color in the farmers' market space (Slocum, 2008).

Moreover, Cadieux and Slocum argue that it is highly important for food scholars to make a clear distinction between food justice and advocacy for an equitable food system (Cadieux and Slocum, 2015). Situated within a larger web of connections, the authors identify four nodes of food justice: trauma and inequity, exchange, land, and labor (Cadieux and Slocum, 2015). Cadieux and Slocum state that "trauma and inequity" as a node of food justice "recognizes structural relations of power as necessary to confront race, class and gender privilege" and "acknowledges the historical, collective traumas, and remembers that the history and expression of trauma varies locally and is fueled by the power of global hierarchies of privilege" (Cadieux and Slocum, 2015, p. 14). The trauma and inequity node "enacts policies that repair the past injustices and trauma that are still felt today" (Cadieux and Slocum, 2015, p. 14). Related to both embodiment geography and the integral aspects of the food justice movement, it is important to validate the linkages these themes, which allows those who are already privileged to achieve even higher status by virtue of their bodies and food-purchasing habits" (Guthman, 2011, p. 193). The results of this study indicate that incentivizing policies have the potential to improve access to fresh produce.

Background

History of food assistance in the United States

The SNAP is the largest federal food assistance program. As an entitlement program, it is implied that the government has committed to supplying the necessary funds required to provide benefits (Wilde, 2013, p. 202). The first Food Stamp Program began in 1939 as a method of disposing agricultural surpluses by selling small vouchers as relief to purchase food (Wilde, 2013). Over the decades, many changes were made, and in 1988, the Hunger Prevention Act was signed into law, and with it began a pilot program for Electronic Benefit Transfer (EBT). In the 1990s, benefits began to be distributed through EBT cards, similar to debit cards. In 2008, the Food Stamp Program was renamed the SNAP to both strengthen the identity of the program as a nutrition program, and recognize the use of EBT rather than physical food stamps (Wilde, 2013). SNAP is tied to the history of industrial agricultural and surplus as its legislation is housed within the United States Department of Agriculture Farm Bill, legislature that dictates agricultural trade and production and is reviewed every five years.

One's eligibility depends on household size and monthly income before taxes. Eligibility for seniors and disabled individuals is determined after taxes and deductions, such as social security or medical costs. The income eligibility guidelines for 2020 are outlined in Table 10.1 (Second Harvest Heartland, 2020). If an individual or household meets the eligibility guidelines, they are able to apply through their local county Department of Human Services.

How to use SNAP/EBT at a farmers' market

In order to use SNAP at a farmers' market, there are a few steps. Most markets use a token system. An EBT user makes their transaction at the customer service or information tent. The shopper hands over their Minnesota EBT card to a volunteer

Table 10.1 SNAP Eligibility Guidelines (US Dollars)

	General Households	*Seniors (60+) and Disabled*
Household Size	*Monthly income before taxes*	*Monthly income after taxes (deductions may be applied)*
1	$1,718	$1,041
2	$2,326	$1,410
3	$2,933	$1,778
4	$3,541	$2,146
5	$4,149	$2,515
Each additional member, add	$608	$369

Source: Second Harvest Heartland 2020

or manager, who asks if they are using food or cash benefits. When shoppers apply for SNAP, which is dependent on household size and income, some receive food benefits and some receive cash benefits. The manager or volunteer asks how much money they would like to take out, then makes the transaction and hands the shopper the corresponding number of tokens, which are each worth a dollar. If the shoppers spend up to ten dollars in SNAP, they are eligible for a ten dollar match through Hunger Solutions MN. For example, if the shopper spends ten dollars, they are given ten market bucks cards to spend like cash, resulting in 20 dollars to spend. This small dollar amount may incentivize smaller, more frequent trips.

Methods

For this study, we selected three farmers' markets within Minneapolis and St. Paul (Figure 10.1). Farmers' markets chosen for the study needed to have a few elements: (1) an existing EBT acceptance program, (2) a busy shopping environment, and (3) managers with the willingness to allow the research to be conducted. There are seven farmers' markets in the Twin Cities that accept SNAP/EBT, of which we have chosen three (Minnesota Department of Human Services). The first market site, the St. Paul Farmers' Market, is located in downtown St. Paul in the Lowertown neighborhood. The second market site, Midtown Farmers' Market, is located in the Longfellow neighborhood of Minneapolis. The third market site, Minneapolis Farmers' Market, is located in the North Lyndale neighborhood in Minneapolis. These three markets were chosen because they each serve a distinct geographic area of the city with little overlap. Additionally, each market is well established in the metro area and in their respective neighborhoods. Both Minneapolis Farmers' Market and St. Paul Farmers' Markets operate smaller market sites throughout the week, but this study focused solely on their respective primary locations (Alhadeff, 2020).

For five consecutive weeks, the first author attended each market. On Tuesday afternoons (3–7 p.m.) or Saturday mornings (9–1 p.m.), the Midtown Farmers' Market was targeted. The Minneapolis Farmers' Market was targeted on Friday

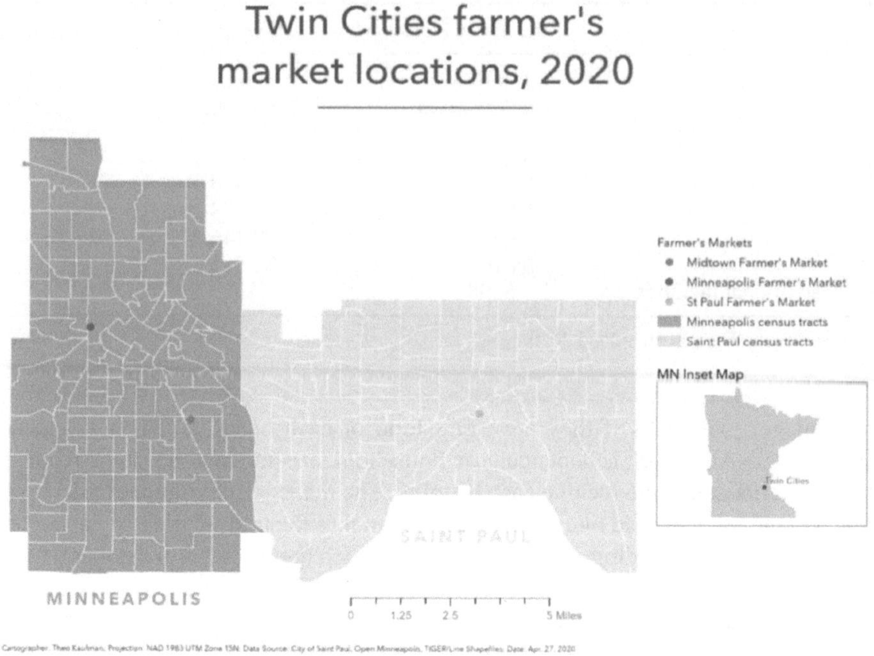

Figure 10.1 Twin Cities Farmer's Market's Locations, 2020

mornings (9–12 p.m.), and the St. Paul Farmers' Markets on Sunday mornings (9–12 p.m.). Over the course of the five weeks, surveys were distributed in English, Spanish, or Hmong to shoppers using EBT. In order for a shopper to use their EBT card, a shopper has to pay for tokens that are able to be used like cash at that particular market. Tokens are exchanged in one dollar increments and never expire. With the permission of the market managers, the first author sat at the information tent alongside the managers. When a shopper used their EBT card, the researcher would ask if they would like to take a short survey. At each of the markets, the majority of the shoppers agreed to participate. Interviews were also conducted with four key informants. Approximately 70 surveys were collected at three sites. Participants were asked basic questions regarding how frequently they shop at the market, how far they travel, how long they have been using SNAP, how long they have shopped at the market, and what proportion of their income is spent on food and covered by SNAP (Alhadeff, 2020). Sample bias may exist due to language and literacy abilities of the participants, despite the three translations in Spanish, Somali, and Hmong. Moreover, toward the third and fourth weeks of my study, the same EBT shoppers reappeared, suggesting a decrease in the number of new shoppers. No explicit questions were asked regarding demographic information, such as race, gender, or age. Despite this, observations were made about the gendered patterns of shoppers and participants in the study.

Findings

We describe the results of the survey data[3] from each market and then compare the results that describe trends of the limits of transportation access as they impact the ability of shoppers to attend a given market. In this analysis, we analyze the survey responses for seven questions. The questions are as follows:

Table 10.2 Survey Questions

1. What mode of transit do you take to the market? (Car, Bus, Walk, Light Rail, Bike)
2. How long is your commute to the market? (less than 10 minutes, 15 minutes, 15–30 minutes, 30 minutes – 1 hour, 1 hour +)
3. How often do you shop at the market? (1 time per week, 2 times per week, 1 time per month, 2 times per month)
4. How long have you received SNAP? (less than a year, 1–3 years, 3–6 years, 6–10 years, 10 year +)
5. What fraction of your monthly income do you spend on food? (more than 75%, 75%, 50%, 25%, less than 25%)
6. How much of your monthly food budget is covered by SNAP (or other food programs) in comparison to your personal income? (100%, 75%, 50%, 25%, less than 25%)
7. Do you see the Minnesota Hunger Solutions Market Bucks Program? (yes, no)

The scope of this research aims to address the gaps in the current academic literature and promote financial equity at farmers' markets in the Twin Cities. Three gaps were addressed previously: the lack of mixed-methods research, the lack of a range of scale, and the lack of communication between scholars and non-profit organizations in the food policy and food justice fields. This study applies a mixed-methods approach and makes policy recommendations to place major actors into conversations.

The results of this study, both quantitative and qualitative, are discussed as they relate to existing literature and the ongoing scholarly conversations regarding social barriers to farmers' market spaces. Connections are drawn between the embodiment geography literature and the patterns that we found in the study.

Minneapolis farmers' market

Located in the North Loop neighborhood, Minneapolis Farmers' Market has served its community since 1937. With two smaller locations in the Twin Cities, the market includes 170 stalls and over 200 vendors. The market has three permanent structures that function as both a covered walkway for shoppers and a space for vendors to sell their products. In comparison to the other two markets in this study, this market has a small kiosk where shoppers can buy chips, donuts, coffee, bottled water, and soda at inexpensive prices. The market is accessible by the Blue Line and Green Line light rail routes at the Target Center station with a 13-minute

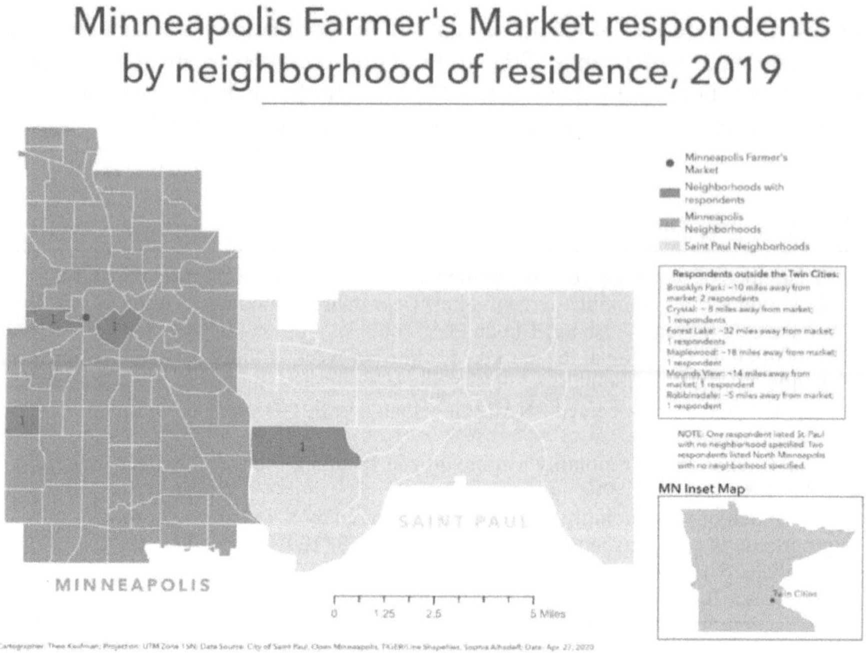

Figure 10.2 Minneapolis Farmer's Markets Respondents by Neighborhood of Residence, 2019

walk. Minneapolis Farmers' Market is open from May through October. Most of the days the first author spent at the Minneapolis Farmers' Market were quiet. Vendors tended to talk among themselves and based on observations, shoppers seemed to walk around, pick out what they wanted to buy, and then leave. In comparison to the other two market sites, there were far fewer people who spent less time overall. At Minneapolis Farmers' Market, there were 14 survey respondents (Figure 10.2).

Midtown farmers' market

Established in 2003, Midtown Farmers' Market was founded by the Corcoran Neighborhood Organization with the goal of creating a public market at the transportation center at the Hiawatha light rail station. Since May 2019, the market has temporarily relocated to Longfellow while there is construction at their permanent site. Midtown Farmers' Market is accessible via the Blue Line, the 7, 21, and 53 bus, and MN Highway 55. During the Tuesdays and Saturdays that I conducted research, the market was bustling. Live music from a local brass band or an acoustic duo filled the setting as shoppers perused, picking out fresh fruits and vegetables, cheeses, meats, and crafts. Interestingly, by my observation, there were more shoppers with physical disabilities in comparison with the other two markets. On

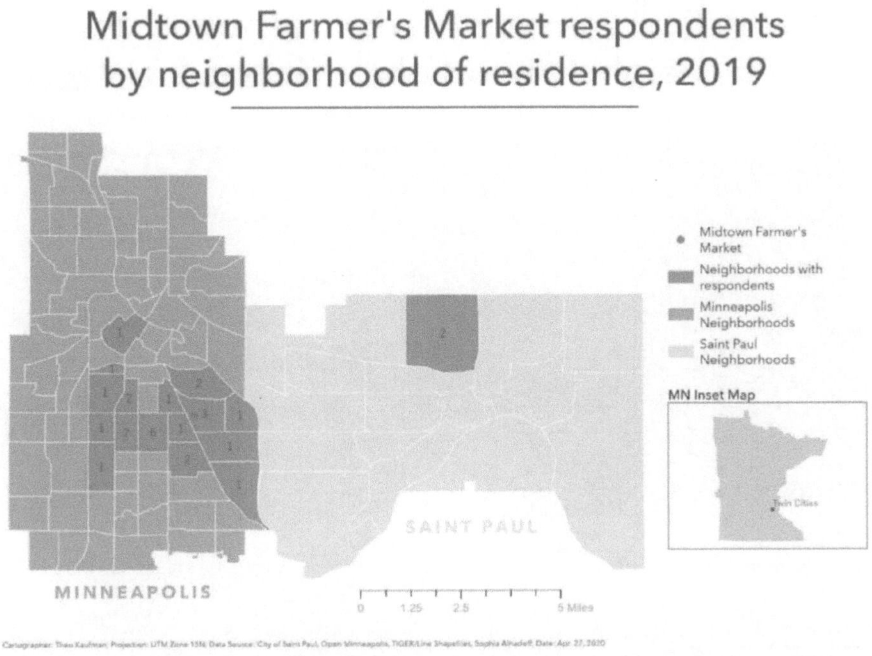

Figure 10.3 Midtown Farmer's Markets Respondents by Neighborhood of Residence, 2019

Saturday mornings, in the tent next to the information booth, community tables were set up to share information about community arts events with shoppers. In comparison to the other two markets in the study, Midtown Farmers' Market had the most sense of community and politically liberal tone. At Midtown Farmers' Market, there were 32 survey respondents (Figure 10.3) (Alhadeff, 2020).

St. Paul Farmers' Market

Located in the Lowertown neighborhood on Broadway between 4th and 5th streets, St. Paul Farmers' Market is the oldest farmers' market in the Twin Cities, opening in 1852. Today, the market is open all year round and has 167 open air stalls. St. Paul Farmers' Market is accessible via multiple bus routes (21, 54, 63, 70, 94, 353, 417, 480, 484, 489) as well as I-94 and MN highway 52. On the St. Paul Farmers' Market website, the market provides information about how to use EBT, including how to exchange EBT for tokens, and which items can be purchased using EBT.[4] During my research, the market was fairly busy with many shoppers walking around the open air environment. On Sunday mornings, there was a raffle for two sports tickets. Based on my observation, there were more young, white families in comparison to the other two markets. My observation was that the market employees were less friendly and inviting toward EBT

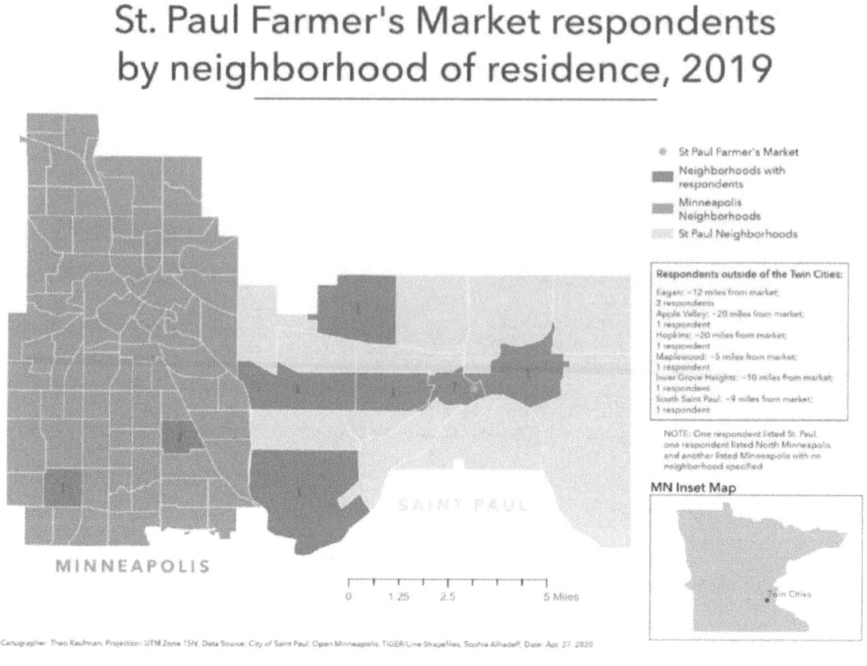

Figure 10.4 St. Paul's Farmer's Markets Respondents by Neighborhood of Residence, 2019

shoppers than the other two markets. At St. Paul Farmers' Market, there were 26 survey respondents (Figure 10.4) (Alhadeff, 2020).

SNAP usage

Three questions were asked to identify how SNAP benefits factor into food purchasing decisions. The survey questions were designed on the basis of what is in the literature. The questions were as follows: What fraction of your monthly income do you spend on food? How much of your monthly food budget is covered by SNAP (or other food programs) in comparison to your monthly income? How long have you received SNAP?

When asked "how long have you received SNAP?" the majority of participants indicated that they have used SNAP for either less than one year or 1–3 years. At Minneapolis, 50% of shoppers have received SNAP for 1–3 years, followed by 21.4% have received benefits for 3–6 years. In contrast, at St. Paul Farmers' Market, 22.7% of participants have received SNAP for less than a year, followed by 31.8% for 3–6 years, and 27.3% for 1–3 years. At Midtown Farmers' Market, there is a slightly more even distribution. 32.3% of respondents have received benefits for 1–3 years, 25.8% for 3–6 years, and 22.6% for less than a year.

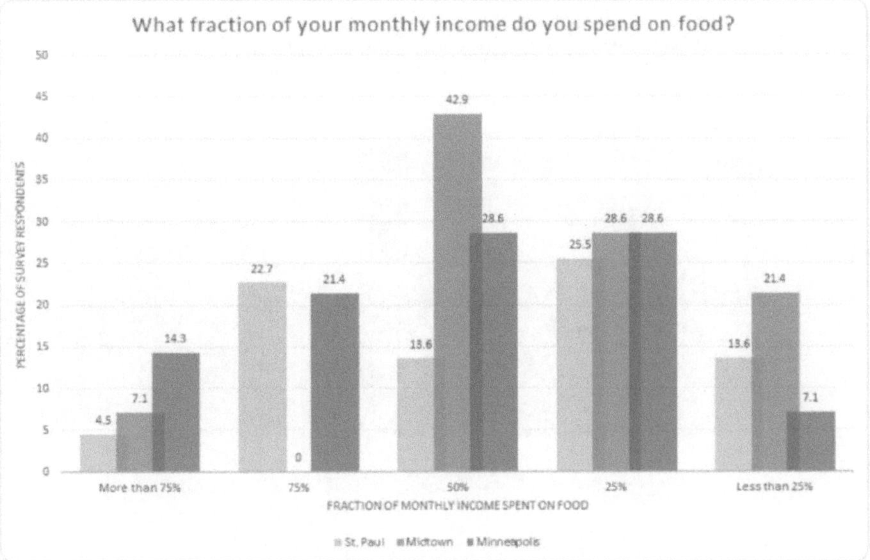

Figure 10.5 What fraction of your monthly income do you spend on food?

Interestingly, at Midtown Farmers' Market, 9.7% of respondents have received SNAP for more than 10 years.

When asked "what fraction of your monthly income do you spend on food?" the majority of respondents selected less than 25%, 25%, or 50% (Figure 10.5). At Midtown Farmers' Market 42.9% of participants spend 50% of their monthly income on food. At both Midtown Farmers' Market and Minneapolis Farmers' Market, 28.6% of respondents spend 25% of their monthly income on food, whereas 25.5% of respondents at St. Paul Farmers' Market spend 25% of their monthly income on food. Interestingly, a higher percentage of respondents at Minneapolis Farmers' Market spend more of their monthly income on food. 14.3% of respondents spend more than 75% of their monthly income on food, and 21.4% spend 75% of their monthly income on food. This trend differs greatly from Midtown Farmers' Market, where only 7.1% of respondents spend more than 75% of their monthly income on food, and 0% of respondents spend 75% of their monthly income on food. The large proportion of respondents reporting spending more than 75% of their income on food could be due to those shoppers receiving other federal assistance, such as energy assistance, disability assistance, or social security.

The results for the responses to this question "How much of your monthly food budget is covered by SNAP or other food programs in comparison to your personal income?" differ widely based on market location (Figure 10.6). At St. Paul Farmers' Market, 31.8% of respondents report SNAP covering 50% of their food budget, 27.3% of respondents report SNAP covering 75% of their food budget,

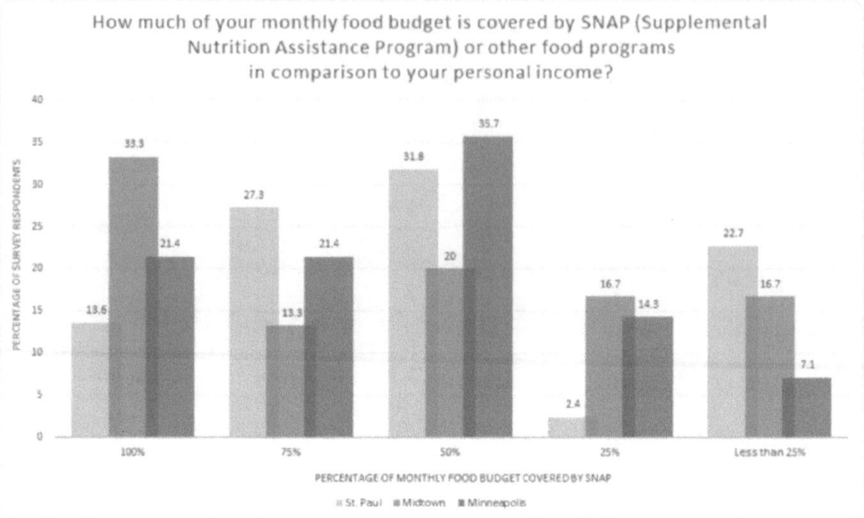

Figure 10.6 How much of your monthly food budget is covered by SNAP or other food programs in comparison to your personal income?

and 13.6% of respondents report SNAP covering 100% of the food budget, while only 2.4% of respondents report SNAP covering 25% of their food budget, and 22.7% of respondents report SNAP covering less than 25% of their food budget. Minneapolis Farmers' Market follows a similar trend. At Minneapolis Farmers' Market, 35.7% of respondents report SNAP covering 50% of their food budget, 21.4% of respondents report SNAP covering both 100% and 75% of their food budget, while only 14.3% of respondents report SNAP covering 25% of their food budget, 7.1% of respondents report SNAP covering less than 25% of their food budget. At Midtown Farmers' Market, 33.3% of respondents report SNAP covering 100% of their food budget, 20% of respondents report SNAP covering 50% of their food budget, 16.7% of respondents report SNAP covering both 25% and less than 25% of their food budget, and only 13.3% of respondents report SNAP covering 75% of their food budget.

The purpose of asking the question "Do you use the Minnesota Hunger Solutions Market Bucks Match Program" was to assess the usage and understanding of the Minnesota Hunger Solutions Market Bucks Match Program (Figure 10.7). The Market Bucks Match Program allows SNAP recipients to receive ten additional dollars to spend at the farmers market when ten dollars are spent with an EBT card. During all of the transactions the first author witnessed during the research, market managers gave out Market Bucks. At Midtown Farmers' Market, 90.6% of survey participants responded by saying that they use this program. This indicates that the overwhelming majority of shoppers at this market know

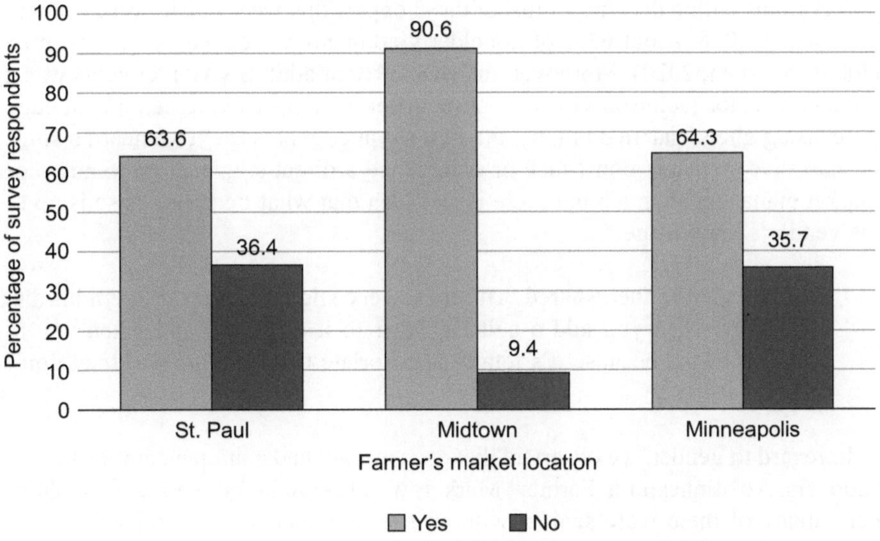

Figure 10.7 Do you use the Minnesota Hunger Solutions Market Bucks Match Program?

about and understand how the program works. In contrast, 63.6% of survey participants responded saying that they use this program at St. Paul Farmers' Market and 64.3% of survey participants responded saying that they use this program at Minneapolis Farmers' Market, indicating that SNAP recipients at those markets have not received the same amount of information about the program.

As represented in the graphs earlier, most shoppers at all three markets have been using SNAP for 1 to 3 years, with a comparable number of respondents who have used it for less than a year. This is indicative that new SNAP users are aware of the ability to shop with SNAP at the farmers' market, but simultaneously, the disparities between shoppers' usage of the MN hunger solutions market bucks indicate a lack of education on behalf of the market managers.

Social barriers and the environment of the farmers' market

In addition to quantitative results, marketing and the environment that each market creates and maintains greatly impacts accessibility for shoppers. Drawing from Slocum and Cadieux's (2015) discussion of trauma and inequity, and the alienation of people of color by the emergence of market tourism, in interviews with shoppers, we asked questions about their lived experience as an EBT shopper at the farmers' market.

In an attempt to better understand the embodied experience of the shopper, we asked shoppers to recall and share a particular experience, whether positive or negative, of using EBT at the market. Following the trends of the literature and observations within the study, most of the shoppers that we spoke to were women. Notably, in 2018, about 63% of nonelderly adult SNAP recipients were women (Johnson-Green, 2020). Moreover, in 2018, 33% of adult SNAP recipients was a woman of color (Johnson-Green, 2020).[5] After a shopper expressed a challenge purchasing ghee (clarified butter), the first shopper expressed "I've heard a similar narrative – people don't face or experience a direct stigma from vendors or market managers so much but there is this idea that what I can purchase is up to the vendor's knowledge."

> The interviewee then shared "But these very slight differences are a hurdle. Especially when you add a cultural layer to it of course and when you're trying to buy it because it's totally appropriate to be buying and they don't consider it food.

In regard to gender, we observed that each market had a unique composition of shoppers. At Minneapolis Farmer' Market, we observed more single adult shoppers, many of them were seniors with an even number of men and women. At Midtown Farmers' Market, we observed that many of the shoppers were women, almost all of the Somali shoppers were women. Additionally, this market had more queer shoppers than other locations. Lastly, at St. Paul Farmers' Market, we observed more white couples with young children. In order to work toward policy transformations that emphasize gender equality, the intersection of gender, SNAP usage, and the trend of women as caretakers and household shoppers should be noted.

In addition to the general understanding of the intricacies of SNAP, vendors may also have a more conservative approach as they are penalized if they do not obey EBT guidelines. According to the Center for Agriculture and Food Systems' Farmers' Market Legal Toolkit, there are four ways a vendor could be penalized for violating SNAP provisions. First, the Food and Nutrition service may "disqualify the market (or vendor) from the SNAP program on a temporary or permanent basis" (Center for Agriculture and Food Systems, 2020). Second, the Food and Nutrition Service "may issue monetary penalties against the market (or vendor) if FNS determines that disqualification would cause hardship to participating SNAP customers" (Center for Agriculture and Food Systems, 2020). Third, the market's "responsible official" who is "responsible for ensuring the market will comply with the law and FNS regulations, policies, and other guidance on SNAP" could be prohibited from SNAP authorization in the future (Center for Agriculture and Food Systems, 2020; Klisch and Soule, 2020). Lastly "market personnel (or the vendor) may be subject to criminal sanctions for intentional fraud or trafficking (purposefully exchanging ineligible items for SNAP funds)" (Center for Agriculture and Food Systems, 2020; Klisch and Soule, 2020). Due to these four potential outcomes, vendors may be less willing to be flexible with SNAP acceptance.

To connect the quantitative results to the ongoing theoretical discussion, the disparity between shoppers' usage of the Market Bucks Match Program. This lack of usage may be due to a lack of education about the program. Thus, it may also be indicative of attention and resources devoted to SNAP/EBT shoppers. While the decisions of what the shoppers are able to purchase can be restricted on the basis of the information given to vendors, the purchasing power at large is limited if the shopper does not know about or is not offered the market bucks match, which allows for ten additional dollars to spend at the market.

When looking forward to ways to improve the culture of farmers' markets, key actors need to ask to whom they are catering their farmers' markets. Does that audience promote financial equity and an inviting, accessible environment to shop? If not, what are the necessary steps to increase equity and inclusion in these spaces? Drawing from feminist political ecology, markets must recognize social difference, such as race, class, gender, sexuality, and able-bodiedness in their market space and implement policy toward to create a shopping environment that supports the needs of low income shoppers (Grosz, 1994; Shi and Hodges, 2015).

Policy recommendations and conclusion

To best suggest policies that would adequately address the drawbacks of SNAP usage at farmers' markets, we discuss the landscape of food insecurity in the Twin Cities. Drawing from this analysis, we then develop what we define as an ideal food policy with a particular emphasis on farmers' markets. Finally, we transform the needs of the Twin Cities community into next steps for a multi-scalar set of policies.

Key actors

When proposing potential policy solutions to alleviate transportation and social barriers to farmers' markets, all of the key actors that facilitate and maintain a farmers' market operation must be addressed. Farmers' markets include both inward and outward facing actors. Inward facing actors include managers, volunteers, donors, along with first time and current EBT shoppers. Outward facing actors include Hunger Solutions Minnesota and the Twin Cities Metropolitan Council. Both inward and outward facing actors will need to be addressed when proposing policy solutions,

Despite geographic proximity, Minneapolis and St. Paul operate in separate legislative bodies and therefore have different policies. Both counties are connected to the Metro Food Access Network (MFAN) which "is a network sponsored by State Housing Initiatives Partnerships grant funding from Ramsey, Dakota, and Hennepin County Public Health Departments." Both Hennepin and Ramsey County have fairly comparable councils working to combat food insecurity, which work on issues in both Minneapolis and St. Paul. In Minneapolis, the Minneapolis Food Action Plan was developed in 2012 and strives to develop a "roadmap toward a more equitable, climate resilient, just and sustainable local food system and local food economy" with seven guiding principles of which (2) Social Determinants of Health, (3) Recognition, Reparations and Respect, and (4) Food Access are included (City of Minneapolis, 2020). The Ramsey

County Food and Nutrition Commission functions as its counterpart (Ramsay County, 2020).

Effective food policy in the Twin Cities and steps for implementation

In order to craft effective food policy in the Twin Cities, a multi-scalar approach combined with an integrated food policy would greatly benefit the food insecure population. Food policy in the Twin Cities must be inclusive, integrated, and accessible. Given the function of the Minneapolis Food Action Plan and the Ramsey County Food and Nutrition Commission, it is evident that both cities are making efforts to address food insecurity. One possibility to improve the integration of food policy is a partnership with the Metropolitan Council. The goal of the Metropolitan Council is "to foster efficient and economic growth for a prosperous metropolitan region" (Metropolitan Council, 2019). Creating a network between these three organizations that would delineate roles and responsibilities would provide Minneapolis and St. Paul a structure to meet the goals of an effective food policy, as defined previously. Moreover, by establishing a central governing body, the Metropolitan Council would then have the potential to coordinate with other organizations and farmers' markets. This proposed regional approach would function more successfully than the current county approach as pooled resources and a broader understanding of the intricacies of the wider landscape of food policy, lending itself to cooperative implementation. This policy design holds the potential to increase collaboration between large actors, such as governing bodies, and small actors such as food shelves and farmers markets.

The second key actor who should be part of the policy solution is Hunger Solutions Minnesota, a nonprofit organization that works within the state and funds the market bucks match program. The mission of the organization is to "take action to assure food security for all Minnesotans by supporting programs and agencies that provide food to those in need, advancing sound public policy, and guiding grassroots advocacy" (Hunger Solutions, 2021). With particular attention to "guiding grassroots advocacy,' we encourage Hunger Solutions MN to write an integrated training manual to distribute to all of the Farmers" Markets in the cities and the state more broadly that accept EBT. This manual would streamline the information needed to train market volunteers, vendors, or managers on the intricacies of SNAP acceptance at the market. A streamlined training manual would clarify the rules of accepting SNAP and take part of the responsibility off of the market staff to train vendors and volunteers. Additionally, Hunger Solutions should create a simple, legible graphic describing how to use EBT and what you can buy at the market. This infographic should be translated into Spanish, Somali, and Hmong.

Farmers' markets: a means of assessing nutritional food

In the context of the Twin Cities, farmers' markets should be included when forming effective food policy. Building from the regional planning approach through an organization such as Metropolitan Council, the council should create a section of regional policy that focuses specifically on the needs of farmers' markets in regard to both vendors and shoppers, with particular attention to low-income communities and the acceptance of SNAP. The inclusion of small actors would allow for a necessary perspective when defining an effective food assistance policy that works toward combating food insecurity in the Twin Cities. Moreover, working with Hunger Solutions to create an integrated training manual for internal actors as well as an infographic for shoppers would allow for a more inclusive environment.

Policies can also be implemented at the state and federal levels to improve the effectiveness of SNAP usage at farmers' markets across the state of Minnesota and the United States. Drawing from political ecology, which outlines how local outcomes to situations are directly influenced by decisions and policies enacted at larger scales, proposing policies at multiple scales can help to enhance both physical and social accessibility to farmers' market spaces. The first step to making farmers' markets more inclusive for EBT shoppers would be to implement a policy to provide every farmers' market, regardless of funds or infrastructure, the ability to accept EBT. Resources would include an EBT card reader and tokens, along with a training manual as discussed above. This manual would also include the rules and regulations for vendors as well as shoppers. The second policy recommendation is to create a similar market bucks match program in each state. Similar programs exist in other states, but through federal policy, a financial incentive could be implemented to stretch the value of the SNAP dollar. Next, we would encourage the USDA and county governments across the country to promote farmers' markets as viable retail food outlets. By including farmers' markets in educational material on the federal level, more low-income shoppers would feel invited into those spaces. Lastly, if SNAP benefit eligibility were evaluated in conjunction with the cost of living in a particular location, in theory, individuals and families living in more expensive environments would receive more benefits, allowing for more potential to spend money at Farmers' Markets. Moreover, on the federal level, policies should be developed to clearly explain how to use an EBT card and what can be purchased with SNAP in a variety of languages or an infographic both online and at grocery stores. Together, if developed further, these three policies have the potential to improve the effectiveness of SNAP usage at farmers' markets across the country.

It is crucial that more effective food assistance policy is enacted in the US in order to make progress toward SDG #2 on ending hunger and improving nutrition, and SDG#5 on advancing gender equality and empowering women by 2030. Our study findings suggest that there is a need to improve the inclusivity of farmers' market spaces as well as the built environment to improve the effectiveness of food assistance usage at farmers' markets in the Twin Cities. Thus, we suggest that a critical step to dismantling social barriers to accessing farmers' markets would be an integrated SNAP/EBT training manual with an emphasis on race and gender for managers and volunteers, as well as a clear and translated infographic for shoppers.

Notes

1 This study employs the definition of food insecurity as "lack of consistent access to enough food for an active, healthy life" (USDA, 2019).
2 The results shown in the table and represented in the charts are ratio level data. I converted the raw values to ratio data for each question as there were different numbers of participants in each location. As there were varying numbers of survey respondents at each market, the data represent the percentage of respondents of the total number at each market.
3 Note: During my five weeks of data collecting, the EBT card reader at STPFM was down for three weeks. While the first author was still able to talk with and survey SNAP participants, the total number of shoppers over the course of those three weeks could be lower than normal. Even if a shopper was not able to use their EBT that particular day, most shoppers were willing to take my survey. Additionally, the market managers gave each SNAP participant five extra market bucks to use at the market due to the dysfunctional EBT reader.
4 Once an individual qualifies for the Supplemental Nutrition Assistance Program (SNAP) or cash benefits (Temporary Assistance) they receive an EBT card, also known as a Common Benefit Identification Card (CBIC). The EBT card looks like a debit card. The EBT card allows an individual to buy groceries and other items with your cash benefits at participating stores and other location.
5 While at Midtown Farmers' market, I had a conversation with a shopper about the inadequate information about what can be purchased with an EBT card generally, not just in the farmers' market context. I did not conduct a formal interview with this shopper, so I am not able to cite direct citations.

References

Alhadeff, S. (2020). Canvas Totes and Plastic Bags: The Political Ecology of Food Assistance Effectiveness at Farmers' Markets in the Twin Cities. *Macalester Honors Projects*.

Cadieux, K., Slocum, R. (2015). *What Does It Mean to Do Food Justice?* Minneapolis: University of Minnesota.

Center for Agriculture and Food Systems. (2020). Enforcement and penalties. *Vermont Law School*. https://www.vermontlaw.edu/academics/centers-and-programs/center-for-agriculture-and-food-systems. Accessed on 29 April 2020.

City of Minneapolis. (2020). Sustainability. City of Minneapolis. https://www2.minneapolismn.gov/government/departments/coordinator/sustainability/

Clifford, N., Holloway, S., Rice, S. (2010). *Key Concepts in Geography*. Newbury Park: Sage Press.

Fessler, P. (2019, July 23). *3 Million Could Lose Food Stamp Benefits under Trump Administration Proposal*. https://www.npr.org/2019/07/23/744451246/3-million-could-lose-food-stamp-benefits-under-trump-administration-proposal. Accessed on 15 September 2019.

Gottlieb, R., Joshi, A. (2010). *Food Justice*. Boston: MIT Press.

Grosz, E. (1994). *Volatile Bodies*. Bloomington: Indiana University Press.

Guthman, J. (2011). *Weighing In: Obesity, Justice and the Limits of Capitalism*. Los Angeles: University of California Press.

HLPE. (2020). Food Security and Nutrition: Building A Global Narrative towards 2030. Report #15. High Level Panel of Experts (HLPE), UN Committee on World Food Security (CFS). June. www.fao.org/3/ca9731en/ca9731en.pdf.

Hunger Solutions. (2021). Winning Minnesota's Food Fight. https://www.hungersolutions.org/hunger-data/

International Labour Organization (ILO). (2017). *World Employment and Social Outlook: Trends for Women 2017*. Geneva: ILO. www.ilo.org/global/research/global-reports/weso/trends-for-women2017/WCMS_557245/lang-en/index.htm

Johnson-Green. (2020). Gender and racial justice in SNAP. *National Women's Law Center*. https://nwlc.org/wp-content/uploads/2020/10/Gender-and-Racial-Justice-in-SNAP.pdf

Klisch, S., Soule, K. (2020). Farmers Markets: Working with community partners to provide essential services during COVID-19. *Journal of Agriculture, Food Systems, and Community Development*, *9*(4), 1–5.

Larson, J., Moseley, W. (2012). Reaching the limits: A geographic approach for understanding food security and household hunger mitigation strategies in Minneapolis-St. Paul, USA. *GeoJournal*. 77(1), 1–12.

Metropolitan Council. (2019). *Feeding the Hungry, Serving the Poor*. St. Paul, MN: Met Council.

Moseley, W.G. (2018). Geography and engagement with U.N. Development Goals: Rethinking development or perpetuating the status quo? *Dialogues in Human Geography*. 8(2), 201–205.

Ramsay County. (2020). *Food and Nutrition Commission*. Saint Paul, MN: Ramsay County. https://www.ramseycounty.us/your-government/leadership/advisory-boards-committees/food-and-nutrition-commission

Rao, N. (2020). The achievement of food and nutrition security in South Asia is deeply gendered. *Nature Food*, 1, 206–209.

Rocheleau, D., Thomas-Slayter, B., Wangari, E. (1996). *Feminist Political Ecology: Global Issues and Local Experiences*. London: Routledge.

Second Harvest Heartland. (2020). *The Supplemental Nutrition Assistance Program Outreach in Schools*. Brooklyn Park, MN: Second Harvest Heartland. https://www.2harvest.org/who--how-we-help/services-and-programs/snap-outreach-in-schools.html#.YSfVSN8pDcs

Shannon, J. (2014). What does SNAP benefit usage tell us about food access in low-income neighborhoods? *Social Science in Medicine*. 107, 89–99.

Shannon, J. (2016). Beyond the supermarket solution: Linking food deserts, neighborhood context and everyday mobility. *Annals of the American Association of Geographers*. 106(1), 186–202.

Shi, R., Hodges, A. (2015). Shopping at farmers' markets: Does ease of access really matter? *Renewable Agriculture and Food Systems*. 31(5), 441–451.

Slocum, R. (2007). Whiteness, space and alternative food practice. *Geoforum*. 38(3), 520–533.

Slocum, R. (2008). Thinking race through corporeal feminist theory: Divisions and intimacies at the Minneapolis Farmers' Market. *Social & Cultural Geography*. 9(8), 849–869.

Slocum, R., Cadieux, K. V. (2015) Notes on the practice of food justice in the US: understanding and confronting trauma and inequity. *Journal of Political Ecology*. 22, 27.

Stephenson, G., Lev, L., Brewer, L. (2008). 'I'm getting desperate': What we know about farmers' markets that fail. *Renewable Agriculture and Food Systems*. 23(3), 188–199.

United Nations (UN). (2016). *Transforming Our World: The 2030 Agenda for Sustainable Development*. New York. United Nations. https://sustainabledevelopment.un.org/

content/documents/21252030%20Agenda%20for%20Sustainable%20Development%20web.pdf
US Department of Agriculture. (2019). *Definitions of Food Security.*
Wilde, P. (2013). *Food Policy in the United States: An Introduction.* New York: Routledge.
Winne, M. (2008). *Closing the Food Gap: Resetting the Table in the Land of Plenty.* Boston: Beacon Press.

Part 3

Target 5.6

Universal access to reproductive health and rights

This third section addresses *Target 5.6*, and the two *Indicators* making up this target (*5.6.1: Proportion of women aged 15–49 years who make their own informed decisions regarding sexual relations, contraceptive use and reproductive health care*, and *5.6.2: Number of countries with laws and regulations that guarantee women aged 15–49 years access to sexual and reproductive health care, information and education*). The four chapters making up this section highlight the importance of women being able to make their own informed decisions regarding reproductive health, and the need for the access to public health education, health promotion, and associated healthcare services. The geographical focus is SSA (Ghana) and Asia (Cambodia). The four chapters making up this section illustrate the interconnectivity of *Goal 5* and Target 5.A (see Introductory Chapter 1, Figure 2), with many other SDGs, including *Goal 1: No Poverty, Goal 2: Zero Hunger, Goal 3: Good Health and Well-Being, Goal 8: Decent Work and Economic Growth, Goal 10: Reduced Inequalities*, and *Goal 12: Responsible Consumption and Production.*

DOI: 10.4324/9780367743918-13

11 Internal migration as a determinant of antenatal care in the Brong-Ahafo Region, Ghana

Does length of residence matter?

Jemima Nomunume Baada and Bipasha Baruah

Introduction

The persistent high rates of maternal and child mortality in low- and middle-income contexts, including Ghana, remain a public health concern. In 2017, SSA recorded 542 maternal deaths per 100,000 live births, compared to the global average of 211 per 100,000 (WHO et al., 2019). The same trend is reported for under-5 child mortality, for which more than half of the 5.2 million global reported deaths occurred in SSA (UNICEF, 2020). Although Ghana has worked to reduce maternal deaths from 760 per 100,000 in 1990 to 308 per 100,000 live births as of 2017, the country's maternal, under-5, infant and neonatal mortality figures are still alarming relative to the global averages. For instance, Ghana's under-5 mortality currently stands at 52 deaths per 1,000 live births, compared to 75.8 for SSA and 37.7 globally. The country's infant mortality rate is also 37, versus 51.7 in SSA and 28.2 across the world. Finally, Ghana's neonatal rates stand at 25 per 1,000 live births, in comparison to 27.5 for SSA and 17.5 for the global average (Statistical Service Accra, 2018; WHO et al., 2019). Despite the country's relative success in addressing maternal and child mortalities compared to the regional (SSA) outcomes, these figures still fall far below the global means.

In response to these high maternal and child mortality rates, the Sustainable Development Goals (SDGs), specifically, targets 3.1 and 3.2, have prioritized reducing maternal and child deaths by more than 50% by 2030. To achieve this, SDG target 3.7 recommends universal access to sexual and reproductive health services, including antenatal care (ANC) for all expectant mothers (UN-SDGs, 2015). According to the World Health Organisation (WHO), ANC refers to "the care provided by skilled health-care professionals to pregnant women and adolescent girls in order to ensure the best health conditions for both mother and baby during pregnancy" (2016, p. 1). Given this urgency and the importance of ANC for achieving SDG goal 3, the WHO's ANC healthcare model also recommends that women visit ANC within the first trimester of pregnancy (gestational age of <12 weeks), and follow up with at least seven visits before delivery (Moller et al., 2017; WHO, 2016).

The importance of timing and early utilization of ANC for lowering mortality rates among women and children, especially in low-and-middle-income contexts,

DOI: 10.4324/9780367743918-14

has been emphasized by scholars (see Atuoye et al., 2017; Heredia-Pi et al., 2016; Kuuire et al., 2017; Razum et al., 2018). For instance, it is observed that early and sufficient use of ANC ensures that pregnant women are able to access information regarding diet and lifestyle options for promoting mother and child well-being, and leads to the early detection of potential complications that may arise during delivery (Arunda et al., 2017; Thangaratinam et al., 2012). Similarly, Kuuire et al. (2017) report that ANC acts as an opportunity for preventing potential causes of and managing existing conditions that could result in maternal and new-born morbidity or mortality. Furthermore, Tekelab et al. (2019), in their systematic review on SSA, reveal that at least one ANC visit with a skilled healthcare provider is associated with a 39% decrease in the risk of neonatal death.

Despite these identified benefits of early timing and effective utilization of ANC for maternal and child health outcomes in low- and middle-income contexts such as Ghana, women's access to these services may be influenced by their socioeconomic and demographic backgrounds (see Arthur, 2012; Heredia-Pi et al., 2016; Kuuire et al., 2017; Moller et al., 2017). In Nigeria, Malawi, and Ghana, Kuuire et al. (2017) and Atuoye et al. (2017) observe that the poorest women were more likely to miss the timing of their first ANC visit, not meet the required number of ANC visits, and not use a skilled birth attendant. Consistent with this, Ganle et al. (2014) found that in Ghana, women without secondary educational attainment were less likely to utilize ANC. Similarly, older women and those with higher birth parity (referring to the number of times that a woman has delivered a fetus of gestational age 24 weeks or older, irrespective of whether this was a live or stillbirth) were less likely to utilize ANC compared to younger women with lower birth parity (Asundep et al., 2014; Owusu and Yeboah, 2018; Pell et al., 2013; Razum et al., 2018).

In addition, women's geographical and locational characteristics tend to define their ease of access to health facilities where they can attend ANC. Yet, regional and urban bias in siting health facilities provides women in resourced regions and in urban areas better access to ANC services, relative to their counterparts in under resourced regions and rural areas (Antabe et al., 2019; Kuuire et al., 2017; Pell et al., 2013). Thus, according to Apanga and Awoonor-Williams (2018) and Atuoye and Luginaah (2017), the relatively high maternal and child mortality rates in rural areas of Ghana arise from poor access to ANC in these settings. Rural areas are also characterized by low numbers of skilled birth attendants, poor infrastructure development, and limited health facilities that directly affect women's access to ANC. While policies, such as the free maternal health insurance enrollment, have been introduced by the government of Ghana to increase women's use of ANC – especially for the poor and underprivileged – expectant mothers in rural areas continue to underutilize reproductive health services due to the lack of health facilities and personnel, and the added transportation costs of needing to travel outside their communities to seek care (Kuuire et al., 2017; Rishworth et al., 2016a).

Research further suggests that barriers to ANC may be amplified for women within the context of migration. In Ghana, Baada et al. (2021) and Owusu and

Yeboah (2018) show that migrant women report poor reproductive health outcomes compared to non-migrant women. This has been attributed to the intersections of their settlement patterns, low socioeconomic status, outsider status, and sociocultural factors, such as language, beliefs, and practices (Asanin and Wilson, 2008; Moller et al., 2017; Owusu and Yeboah, 2018). For migrant women who settle in remote rural areas, meeting their reproductive health needs, including ANC visits, may be particularly challenging due to the distance to health centers, poor transportation services, and a lack of social support in migrant-receiving societies. These outcomes may be further compounded for migrant women due to the gendered labor markets in receiving areas, which often put them at an economic disadvantage (Baada et al., 2019; Owusu and Yeboah, 2018).

Despite the implications of these findings for achieving the SDGs in low- and middle-income contexts, such as Ghana, few studies examine the experiences of ANC use among migrant women in rural areas as a distinct group. The limited studies which explore the reproductive experiences of migrant women often focus on those in urban areas, to the neglect of their counterparts in rural areas (see Owusu and Yeboah, 2018). Furthermore, reproductive health studies on women in rural areas often treat both migrant and non-migrant women as a homogenous group who share similar experiences in accessing ANC (Atuoye et al., 2015; Rishworth, et al., 2016a). This is despite studies which suggest that migration processes, including length of stay, could potentially influence access to and utilization of reproductive health services (Razum et al., 2018). These migration processes may exacerbate the burdens of using ANC in rural destination areas, particularly for recent migrant women. Given the evidence that women's participation in internal rural–rural migration in Ghana has increased substantially (Abdul-Korah, 2011; Baada et al., 2019; Ghana Statistical Service, 2013), it is concerning that there is still a dearth of knowledge about the experiences of ANC use among women in rural migration destinations.

To meet the SDGs on maternal and child mortality in Ghana, it is crucial to understand ANC utilization among migrant women in rural contexts for several reasons. First, population migration is expected to increase as part of ongoing globalization. Climate change trends are also adversely impacting communities in low- and middle-income regions and accentuating population migration (Afifi et al., 2016; Buettner and Muenz, 2020). Second, migration has been identified as a social determinant of health (Dunn and Dyck, 2000), which includes migrant women's reproductive health. Third, the geographical characteristics of rural migration destinations may present unique challenges for migrant women's use of ANC (Arthur, 2012; Kuuire et al., 2017; Baada et al., 2021). Finally, promoting ANC use and resulting health outcomes among migrant women in rural areas relates to eight of the 17 SDGs, including goals three (good health and well-being), five (gender equality), and ten (reduced inequality). Using the case of the growing outmigration of women from rural areas of the Upper West Region (UWR) to the Brong-Ahafo Region (BAR) of Ghana (see Figure 11.1 below), this study seeks to contribute to the migration and health literature by employing the Andersen's Healthcare Utilisation model to examine migrant women's ANC

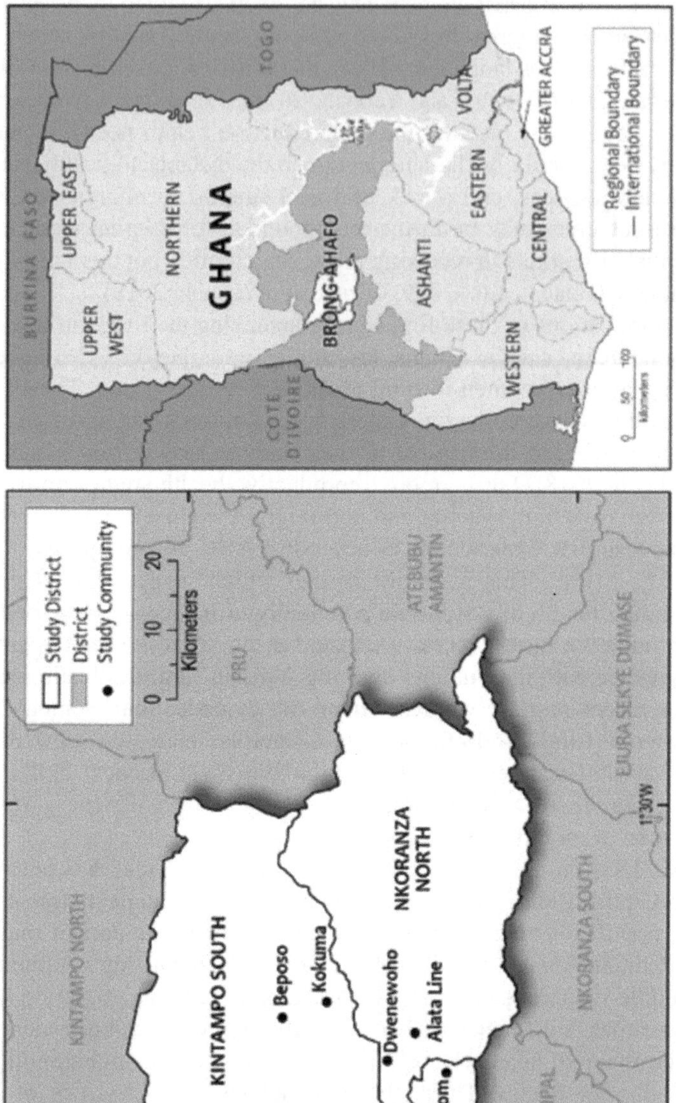

Figure 11.1 Map of BAR showing study areas

Source: Karen VanKerkoerle, Department of Geography, Western University use in rural migration destinations. Our findings are useful for local and international gender and equity-based policies aimed at promoting the health and overall well-being of marginalized groups such as rural migrants, especially women and children.

Andersen's Healthcare Utilisation model

We use Andersen's Healthcare Utilisation model as a theoretical framework for understanding ANC use among women who migrated from the UWR to rural areas of the BAR of Ghana. Earlier versions of this theoretical model (Aday and Andersen, 1981; Andersen, 1968; Andersen and Newman, 1973) have been critiqued for their failure to capture the complexities of social and structural determinants of health, and how these influence health seeking behavior and healthcare utilization. In response, a newer version of the model goes beyond narrow medical access and utilization, to account for the role of other contextual factors, such as familial and social support in shaping health needs, access, and utilization (Andersen, 1995).

The Andersen's Healthcare Utilisation model (1995) proposes that health seeking behavior and successful utilization of health services are shaped by three interrelated conditions: need, predisposing, and enabling factors. Need factors refer to those that require or prompt an individual to seek healthcare. This need may be real – as evaluated from a professional viewpoint, or perceived – as a result of a woman's perceived health risk regarding pregnancy that requires her to seek ANC. For pregnant women, previous experience of delivery is often used as a proxy in determining this need (Amu et al., 2018; Antabe et al., 2020). Need factors are closely tied to predisposing factors, which refer to the demographic and sociocultural characteristics of women preceding ill health. These may include geographical location, ethnicity, age, educational status, and health, religious and cultural beliefs/practices (Amu et al., 2018). For instance, women with higher levels of educational attainment may be more likely to use preventive health services, including ANC, given their access to knowledge about the benefits of such services in averting birth complications (Ochako et al., 2016). Finally, enabling factors are those personal, familial, and communal resources that equip an individual with the means to seek or access healthcare. These enablers include income or wealth status, social support, health insurance, proximity to healthcare services, and access to transportation (Andersen, 1995; Were et al., 2011).

Several studies have applied the Andersen's Healthcare Utilisation model in various health and geographical contexts to understand healthcare utilization, with respect to maternal healthcare services in rural Bangladesh (Amin et al., 2010), intercountry comparison of ANC use in Nigeria and Malawi (Kuuire et al., 2017), and the utilization of breast cancer screening in Kenya (Antabe et al., 2020). Andersen's Healthcare Utilisation model provides a helpful tool for contextualizing migrant women's ANC utilization in rural receiving areas of Ghana, as it can help to unpack how migrant women's sociodemographic characteristics, such as age, ethnicity, and level of education, may potentially affect their ANC use. Moreover, the model is helpful in emphasizing how women's migrant/

outsider status, residence or settlement patterns, and length of stay in destination regions influence their access to and utilization of ANC in rural destination communities of Ghana.

The study context

This study was conducted among women who migrated from the UWR to the BAR for farming purposes. Data were collected from the Nkoranza South, Techiman North and Kintampo South Districts (see Figure 11.1) in the Bono East Region (the current name of the region), which was previously known as the BAR, including at the time of data collection in 2016. Migrations from UWR to BAR have been happening for close to a century now due to colonial and neo-colonial legacies, which have led to uneven development between the northern and southern sectors of Ghana (Songsore, 1979). During British colonial rule, the northern parts of Ghana, which includes the UWR, were excluded from development agenda because they did not fall under the British protectorate. Instead, young men were often recruited from the northern sector as labor for the mines and plantations in southern Ghana (Songsore, 1979). After colonial rule, development policies still prioritized the southern sector as a way of sustaining labor flow to these areas. Although later migrations were no longer forced, the limited economic and educational opportunities in UWR implied that several people had to migrate to the south to access these opportunities (Songsore, 1979). Most north–south migrations in Ghana during this period were male dominated due to cultural factors which discouraged the migration of women (Abdul-Korah, 2011).

From the 1970s onward, however, the migration of women became more common and accepted for several reasons. First, environmental conditions such as rainfall and soil fertility started to deteriorate in UWR. Second, Ghana adopted the Structural Adjustment Programme (SAP), which was supported by the International Monetary Fund (IMF), as a way of addressing the excessive national debts incurred by the country's leaders (Konadu-Agyemang, 2000). Some of the requirements of the SAPs included the removal of agricultural subsidies as a way of reducing state expenditure. These subsidy cuts were particularly marginalizing for people in the UWR, as the majority were poor farmers who depended on state subsidies (Songsore and Denkabe, 1995). These developments, coupled with the existing high rates of poverty and declining climatic conditions, made migration to rural areas of the BAR an alternative livelihood option. Thus, as outmigrations increased, more women (sometimes along with entire families) also began to move out of the UWR (Abdul-Korah, 2011).

The BAR is the destination of choice for several migrants from the UWR due to its biannual rainfall season, relatively better soil fertility, presence of migrant networks due to years of in-migration, and relative proximity to UWR, compared to other southern regions (Abdul-Korah, 2011; Ghana Statistical Service, 2013). Since most migrants from UWR are of low socioeconomic status and migrate for farming purposes, the majority settle in rural communities of BAR due to the relative ease and cost-effectiveness of accessing lands in these areas. These settlement patterns, however, have implications for migrants' health, particularly the reproductive health of migrant women, as these rural communities tend to be remote and lack social infrastructure/amenities including electricity, good roads, health centers, among others (Baada et al., 2021).

Methods and results

Data and analysis

Data for this study were collected from September to December 2016, as part of a master's research project aimed at understanding migrant women farmers' lived experiences in rural parts of the BAR in Ghana. The few prior studies on women's reproductive health in rural areas of Ghana have shown that women generally experience poor health services and outcomes (Atuoye et al., 2015; Rishworth et al., 2016b). Building upon such findings, this study sought to explore migrant women's ANC use, specifically in rural destination areas. For data collection, we first selected the three highest migrant receiving districts in the BAR (i.e., Nkoranza South, Techiman North and Kintampo South districts), based on data from the 2010 Population and Housing Census by the Ghana Statistical Service. A two-staged stratified sampling technique using probability proportional to size framework was employed. First, we grouped communities in the three districts into "rural large towns" and "rural small towns," as these unique locational characteristics tend to shape how women, including migrants, access ANC. In the second stage, we randomly sampled households from both clusters (rural large towns and rural small towns), resulting in the identification and recruitment of 750 households eligible for enumeration. A survey was administered to a total of 700 migrant women aged 18–45 in the three districts, representing a response rate of 93.3%. Our survey instrument was pretested prior to the commencement of data collection for content relevance and clarity. Among others, survey questions asked migrant women about how long they had been resident in the BAR, how often they used ANC, and their experiences of using these services. Ethics approval for this study was obtained from Western University's Non-Medical Research Ethics Board.

Measures

The dependent variable for this study explores whether women attended ANC at least four times when they were pregnant (0=no; 1=yes). Our independent and control variables are informed by the Andersen's Healthcare Utilisation model. Specifically, there are three sets of variables, namely predisposing, enabling, and need variables. For predisposing factors, we included length of residence (0=more than 10 years; 1=10 years or less), age of respondents (0=15–24; 1=25–34; 2=35–44), location of residence (0=urban; 1=rural), level of education (0=no education; 1=primary education; 2=secondary/higher education), religious affiliation (0=Christian; 1=Muslim; 2=traditionalist), and ethnicity (0=Sissala; 1=Waala; 2=Brifo; 3=Dagaaba). We further incorporated one variable each for enabling and need factors, including household wealth (0=richest; 1=richer; 2=middle; 3=poorer; 4=poorest) and the number of births (continuous variable), respectively. Household wealth quintiles were constructed from a composite index based on the household ownership of consumer items, such as drinking water, car, and toilet facilities, among others.

Statistical analysis

We employed two different analyses for this study. For one, we conducted univariate analysis to describe the characteristics of the analytical sample. For another, we relied on regression analysis to understand the association between the dependent (i.e., whether women attended ANC when pregnant) and independent variables (i.e., length of residence). Considering the dichotomous nature of the dependent variable, we used logistic regression analysis (Hosmer et al., 2013). Models were built sequentially. In Model 1, we examined the bivariate association between the dependent and independent variables, and Models 2, 3, and 4 further accounted for predisposing, enabling, and need factors, respectively. Results are shown with odds ratio. Odds ratios larger than 1 imply that women were more likely to attend ANC when they were pregnant, while those smaller than 1 indicate lower odds of doing so.

Results

Table 11.1 shows the findings from the univariate analysis. We found that more than half of women (56%) attended ANC at least four times when they were pregnant. We also found that about one-fourth of women (24%) had lived in the BAR for less than 10 years. The largest age group was 25–34 (40%), followed by 15–24 (32%) and 35–44 (28%). The majority of women lived in rural small towns (82%), did not have any formal education (62%), were Christian (77%), and belonged to the Dagaaba ethnic group (76%). The mean number of given births was 2.55.

Findings from the regression analysis are shown in Table 11.2. In Model 1, we find that the bivariate association between length of residence and ANC attendance is statistically significant. Specifically, women who have been in the BAR for less than 10 years are less likely to attend ANC than their counterparts who have been in the BAR for more than 10 years ($OR = 0.60$, $p < 0.001$). In Model 2, when we accounted for predisposing factors, particularly age of respondents, the statistical significance of the dependent and independent variable was partially reduced ($OR = 0.82$, $p < 0.01$). When we include enabling and need factors in Models 3 and 4, it is particularly highlighted that household wealth plays an important role for attenuating the relationship between length of residence on ANC attendance ($OR = 0.85$, $p < 0.05$).

In addition, we find several control variables to be significantly associated with women's ANC attendance (see Model 4). For example, women who are part of the 25–34 ($OR = 0.57$, $p < 0.05$) and 35–44 ($OR = 0.17$, $p < 0.01$) age categories are less likely to attend ANC than their 15–24 counterparts. In addition, compared to the richest women, their richer ($OR=0.62$, $p<0.05$), middle ($OR=0.53$, $p<0.05$),

Table 11.1 Univariate analysis

	Percentage
Attended ANC at least four times	
No	44
Yes	56
Length of residence	
More than 10 years	76
10 years or less	24
Age of respondents	
15–24	32
25–34	40
35–44	28
Location of residence	
Urban	18
Rural	82
Level of education	
No education	62
Primary education	22
Secondary/higher education	16
Religious affiliation	
Christian	77
Muslim	19
Traditionalist	4
Ethnicity	
Sissala	11
Waala	6
Brifo	7
Dagaaba	76
House wealth	
Richest	19
Richer	21
Middle	19
Poorer	21
Poorest	20
Number of given births†	2.55
Number	441

†Mean reported

poorer (OR = 0.63, $p < 0.05$), and poorest (OR = 0.45, $p < 0.01$) counterparts were all less likely to attend ANC when they were pregnant. Finally, the number of given births (birth parity) was negatively associated with the odds of attending ANC (OR = 0.84, $p < 0.001$), implying that women with higher numbers of given births were less likely to attend ANC than those with lower numbers.

Discussions and conclusion

Despite the progress made over the last few years, Ghana, like most SSA countries, still reports high maternal and child mortality rates relative to the global averages.

Table 11.2 Multivariate analysis predicting use of ANC at least four times during last pregnancy

	Model 1	*Model 2*	*Model 3*	*Model 4*
Length of residence				
More than 10 years	1.00	1.00	1.00	1.00
10 years or less	0.60***	0.82**	0.85*	0.78*
Age of respondents				
15–24		1.00	1.00	1.00
25–34		0.63*	0.61***	0.57*
35–44		0.19***	0.17	0.17**
Location of residence				
Rural large town		1.00	1.00	1.00
Rural small town		0.81	0.88**	0.94
Level of education				
No education		1.00	1.00	1.00
Primary education		0.89	0.95	1.04
Secondary/higher education		0.87	0.92	0.91
Religious affiliation				
Christian		1.00	1.00	1.00
Muslim		0.75	0.76	0.84
Traditionalist		0.45	0.58	0.44
Ethnicity				
Sissala		1.00	1.00	1.00
Waala		0.88	0.82	0.87
Brifo		1.06	0.83	0.75
Dagaaba		0.78	0.70	0.71
House wealth				
Richest			1.00	1.00
Richer			0.57	0.62*
Middle			0.50	0.53*
Poorer			0.61	0.63*
Poorest			0.44**	0.45**
Number of given births				0.84***
Log pseudo-likelihood	−296.01	−209.09	−207.12	−206.54
Wald χ^2	12.66***	91.42***	102.03***	106.69***

*p < 0.05, **p < 0.01, ***p < 0.001; OR = odds ratio; SE = standard error

The country therefore needs to make significant advances in order to meet the SDG targets 3.1 and 3.7 – aimed at reducing maternal mortality to 190 deaths per 100,000 live births and improving child mortality rates by 2030 (UNICEF, 2020; WHO, 2016; World Bank, 2020). Using the case of high rates of internal migration in Ghana, this study contributes to the migration and health literature, and policies aimed at meeting the SDGs, by drawing on the Andersen's Healthcare Utilisation model to examine ANC use among women who migrated from the UWR to rural communities of the BAR of Ghana.

In line with studies (Amin et al., 2010; Atuoye et al., 2017; Baada et al., 2021), which suggest that contextual factors in rural and migration destinations tend to adversely impact women's use of ANC, our findings reveal that migrants' use of ANC may be further affected by their length of stay in rural communities. In the context of this study, migrant women from the UWR who have been in the BAR for less than 10 years are less likely to attend ANC compared to their counterparts who have been in the BAR for more than 10 years. We have two possible explanations for this finding. First, given migrant women's outsider status in receiving societies, recent migrants may have underdeveloped social networks, resulting in poor social capital and support, which are useful for navigating ANC. This finding is consistent with the observation by Story (2014) that women with better social capital are more likely to use ANC and professional healthcare delivery, and have their children immunized. Underscoring the importance of social capital for the use of maternal services, Semali et al. (2015) also found in Tanzania that women with the highest levels of social capital were more likely to deliver in a health facility.

Second, it is also likely that a longer length of stay better positions migrant women to familiarize themselves with the available health facilities and services within and around their communities, and the best ways to access ANC. This finding is also not too surprising, as Asanin and Wilson (2008) show that predisposing and enabling determinants of healthcare, including geographical, sociocultural, and economic factors, could act as barriers that delay migrants' access to appropriate healthcare by as many as nine years. Moreover, our findings support those of Moller et al. (2017), who found that length of stay directly influenced migrant women's ANC use, with a higher coverage (80%) among migrant women who had been in their migration destination five years or longer, compared to those who had been in these destinations two years or less (51%). This finding demonstrates that while all migrant women may face challenges in accessing ANC, recent migrants may require special attention, as their poorly developed social capital and knowledge of the healthcare system in receiving areas makes them particularly vulnerable and presents unique barriers to their effective use of ANC.

Our findings further show that ANC use among migrant women in rural settings in Ghana is influenced by their need factors, including age. Specifically, we found that older migrant women are less likely to attend ANC than their younger counterparts within the 15–24 age group. This finding contrasts those of earlier studies in southern Ghana by Manyeh et al. (2020) and in Nigeria and Malawi by Kuuire et al. (2017), where older women were more likely to use ANC. We again provide two probable explanations for this outcome. First, it is possible that younger migrant women who make up the study participants might have migrated to the BAR as children, and have therefore been residents in the destination area longer, despite their relatively younger ages. Being resident in the destination for a longer period therefore makes them more familiar with the communities and influences their ability to seek ANC. Second, it is likely that in the context of rural–rural migration, younger migrant women experiencing their first pregnancies – in the absence of adequate familial and communal support and coupled with the lack of readily available health facilities/personnel in these rural communities – may perceive their health risks and need as heightened, compared to their more experienced, older counterparts. Furthermore, we find that higher birth parity is associated with a lower likelihood of utilizing ANC. In line with the same observation by earlier scholars (see Antabe et al., 2019; Kuuire et al., 2017; Simkhada et al., 2008), we argue that higher birth parity reduces women's perceived need and predisposition to use ANC, based on their previous experiences of child birth.

Our findings also emphasize the importance of enabling factors such as household wealth on women's use of ANC. For instance, we observe that, compared to the richest women, their richer, middle, poorer, and poorest counterparts are all less likely to attend ANC. This finding is consistent with several studies (see Kuuire et al., 2017; Moller et al., 2017; Rishworth et al., 2016b; Zhao et al., 2012) which demonstrate that socioeconomic determinants such as income, household wealth, and health insurance enrollment all shape women's propensity to use ANC. In the context of our study, this implies that, while the importance of geography to women's ANC use cannot be disputed, the influence of rural settlement patterns on women's barriers to ANC use may be mitigated through the provision of socioeconomic relief packages. This is consistent with research findings which show that financial barriers may affect women's ability, particularly in rural areas, to enroll in health insurance schemes or pay for added healthcare costs that are not covered by insurance (Rishworth et al., 2016b). Furthermore, they support studies which indicate that transportation costs often pose a challenge to women's ability to access maternal healthcare in rural areas (Atuoye et al., 2017). Lastly, as Baada et al. (2021) have shown, women within a rural–rural migration context often experience difficulties in accessing social support, which affects their ability to carve time out of their work schedules to seek ANC. Our study therefore emphasizes that household wealth may cushion migrant women in rural areas by affording them the financial means to meet added healthcare costs, and perhaps enabling them to take time off work to access ANC without worrying about the economic implications.

On the basis of these findings, we make some policy recommendations. First, our study has shown that length of stay is associated with migrant women's use of ANC in rural migration destination areas, which could be attributable to the acquisition of social and economic capital. It is therefore imperative that the local level governments in Ghana develop more direct and proactive outreach programs to provide migrants with information about the services that they can access in their destination communities, including ANC. It is also critical to ensure that these services are accessible, free, and/or as affordable as possible for lower income groups. This could be done through the provision of subsidies in transportation and other healthcare costs for lower income migrant groups. These broader interventions could be supplemented with activities or policies that foster interaction and bonding among migrants and non-migrants in receiving areas. This may be achieved through community and health durbars and sensitization campaigns which bring together migrants and non-migrants. In addition to serving as a means of providing health information and resources to settler communities, these durbars and campaigns could also serve as an avenue for migrants to engage with and develop relationships/trust with non-migrants in receiving communities. These measures could be instrumental in helping recent migrants to develop and nurture enabling resources of healthcare utilization (social and economic capital outlined earlier) sooner, and subsequently increase their ANC use in rural communities.

Furthermore, education aimed at promoting migrant women's understanding of health "need" and preventive healthcare will go a long way to improve ANC use among older migrant women and those with higher birth parity. It is however important to complement these educational strategies with information about how and where to access ANC, as well as avenues for seeking subsidies or financial assistance regarding ANC. These measures would ensure that there is widespread knowledge about ANC within settler communities and encourage accessibility among (older) migrant women experiencing financial and information barriers to these services.

In addition, to achieve the SDGs on maternal and child mortality in Ghana, there is the urgent need for social infrastructural development – particularly the establishment of well-resourced health centers and increased deployment of health workers to these facilities – in rural communities of the country. Finally, our findings emphasize that improving the socioeconomic status of migrant women will better position them to overcome some of the contextual barriers to accessing ANC. For instance, since most rural migrants are farmers, recent migrant women could be assisted with alternative and sustainable livelihood opportunities that keep them economically engaged during the off-farming season. This will help to improve their economic enablers of ANC use such as healthcare costs regarding transportation, insurance, and medication.

Our study is not without limitations. First, the cross-sectional design of the survey implies that our findings are limited to statistical associations, so findings should be interpreted with caution. Furthermore, our data could be subject to recall bias given that migrant women were asked about their last pregnancies. Lastly, we recognize that household wealth may not necessarily reflect migrant women's individual wealth, or their ability to access and utilize this household wealth. As asserted by Baruah (2009), although purely statistical techniques (e.g., household wealth index) may be useful in providing an overview of intrahousehold dynamics, they are limited in their ability to provide a nuanced picture of the subtle negotiations underlying resource utilization between male and female partners within the home. Furthermore, due to these limitations, survey tools may miss the intricacies of resource distribution within the household, which is particularly crucial given that in some cases, increasing household income/wealth may actually lead to women's increased subordination and vulnerability within the household (Baruah, 2009). Given these limitations, it would be helpful for future studies on ANC access and utilization among women in migrant and rural communities to adopt a mixed-methods approach in examining the ways in which intra-household dynamics regarding social, economic, and cultural resources influence women's experiences of ANC.

Nonetheless, our findings contribute to the literature and policy on ANC utilization among migrant women in rural destination contexts, and could be drawn upon by health, development, and policy practitioners to help improve the health and well-being of women, children, and migrant populations in Ghana and similar contexts in SSA. Our findings also contribute toward meeting SDGs three (good health and well-being), and ten (reduced inequality). Finally, addressing ANC challenges among migrant women in rural areas of Ghana will go a long way to help the country meet its gender equality objectives in line with SDG 5 (gender equality). As Brown et al. (2010) emphasize, poor health is one of the main causes of economic deprivation, with resulting effects on overall well-being. Thus, poor antenatal health outcomes will affect migrant women's ability to equitably and effectively participate in economic, social, and political activities, which has implications for bridging gender inequality in Ghana. It is therefore important to address ANC needs among migrant women in rural destination areas to promote their complete physical, mental, and social well-being, and better leverage their participation in diverse spheres of life.

In conclusion, our study on ANC utilization among migrant women in rural areas of Ghana contributes toward attaining the SDGs on gender equality, good health and well-being, and reduced inequalities, particularly among rural and mobile populations. This is crucial, considering the unique social, economic, and cultural vulnerabilities that these groups face. Our findings are also essential and timely, given the need to accelerate action toward meeting the SDGs within SSA and the global south, in order to meet the 2030 agenda. Importantly, worsening climate change effects and increasing globalization will lead to more migrations in SSA and the global south in the years to come. It is thus critical to ensure that the health vulnerabilities of migrant groups, particularly women and children, are mitigated as effectively as possible.

References

Abdul-Korah, G. B. (2011) "Now if you have only sons you are dead": Migration, gender, and family economy in twentieth century northwestern Ghana. *Journal of Asian and African Studies*, 46(4), 390–403. https://doi.org/10.1177/0021909611400016

Aday, L. A., & Andersen, R. M. (1981) Equity of access to medical care: A conceptual and empirical overview. *In Medical Care*, 19, 4–27. Lippincott Williams & Wilkins. https://doi.org/10.2307/3763937

Afifi, T., Milan, A., Etzold, B., Schraven, B., Rademacher-Schulz, C., Sakdapolrak, P., Reif, A., van der Geest, K., & Warner, K. (2016) Human mobility in response to rainfall variability: Opportunities for migration as a successful adaptation strategy in eight case studies. *Migration and Development*, 5(2), 254–274. https://doi.org/10.1080/21632324.2015.1022974

Amin, R., Shah, N. M., & Becker, S. (2010) Socioeconomic factors differentiating maternal and child health-seeking behavior in rural Bangladesh: A cross-sectional analysis. *International Journal for Equity in Health*, 9(1), 9. https://doi.org/10.1186/1475-9276-9-9

Amu, H., Dickson, K. S., Kumi-Kyereme, A., & Maafo Darteh, E. K. (2018) Understanding variations in health insurance coverage in Ghana, Kenya, Nigeria, and Tanzania: Evidence from demographic and health surveys. *PLoS One*, 13(8), 1–14. https://doi.org/10.1371/journal.pone.0201833

Andersen, R. M. (1968) A behavioral model of families' use of health services. *A Behavioral Model of Families' Use of Health Services*, 25.

Andersen, R. M. (1995) Revisiting the behavioral model and access to medical care: Does it matter? *Journal of Health and Social Behavior*, 36(1), 1–10.

Andersen, R. M., & Newman, J. F. (1973) Societal and individual determinants of medical care utilization in the United States. *The Milbank Memorial Fund Quarterly: Health and Society*, 51(1), 95. https://doi.org/10.2307/3349613

Antabe, R., Kansanga, M., Sano, Y., Kyeremeh, E., & Galaa, Y. (2020) Utilization of breast cancer screening in Kenya: What are the determinants? *BMC Health Services Research*, 2, 1–9.

Antabe, R., Sano, Y., Anfaara, F. W., Kansanga, M., Chai, X., & Luginaah, I. (2019) Antenatal care utilization and female genital mutilation in Kenya. *Sexuality and Culture*, 23(3), 705–717. https://doi.org/10.1007/s12119-019-09595-6

Apanga, P. A., & Awoonor-Williams, J. K. (2018) Maternal death in Rural Ghana: A case study in the Upper East Region of Ghana. *Frontiers in Public Health*, 6, 101. https://doi.org/10.3389/fpubh.2018.00101

Arthur, E. (2012) Wealth and antenatal care use: Implications for maternal health care utilisation in Ghana. In *Health Economics Review* (Vol. 2, Issue 1, pp. 1–8). New York, NY: Springer. https://doi.org/10.1186/2191-1991-2-14

Arunda, M., Emmelin, A., & Asamoah, B. O. (2017) Effectiveness of antenatal care services in reducing neonatal mortality in Kenya: Analysis of national survey data. *Global Health Action*, 10(1). https://doi.org/10.1080/16549716.2017.1328796

Asanin, J., & Wilson, K. (2008) "I spent nine years looking for a doctor": Exploring access to health care among immigrants in Mississauga, Ontario, Canada. *Social Science and Medicine*, 66(6), 1271–1283. https://doi.org/10.1016/j.socscimed.2007.11.043

Asundep, N. N., Jolly, P. E., Carson, A., Turpin, C. A., Zhang, K., & Tameru, B. (2014) Antenatal care attendance, a surrogate for pregnancy outcome the case of kumasi, Ghana. *Maternal and Child Health Journal*, 18(5), 1085–1094. https://doi.org/10.1007/s10995-013-1338-2

Atuoye, K. N., Amoyaw, J. A., Kuuire, V. Z., Kangmennaang, J., Boamah, S. A., Vercillo, S., . . ., & Luginaah, I. (2017) Utilisation of skilled birth attendants over time in Nigeria and Malawi. *Global Public Health*, 12(6), 728–743.

Atuoye, K. N., Dixon, J., Rishworth, A., Galaa, S. Z., Boamah, S. A., & Luginaah, I. (2015) Can she make it? Transportation barriers to accessing maternal and child health care services in rural Ghana. *BMC Health Services Research*, 15(1), 1–10. https://doi.org/10.1186/s12913-015-1005-y

Atuoye, K. N., & Luginaah, I. (2017) Food as a social determinant of mental health among household heads in the Upper West Region of Ghana. *Social Science & Medicine*, 180, 170–180. https://doi.org/10.1016/j.socscimed.2017.03.016

Baada, J. N., Baruah, B., & Luginaah, I. (2019) "What we were running from is what we're facing again": Examining the paradox of migration as a livelihood improvement strategy among migrant women farmers in the Brong-Ahafo Region of Ghana. *Migration and Development*, 00(00), 1–24. https://doi.org/10.1080/21632324.2019.1573564

Baada, J. N., Baruah, B., Sano, Y., & Luginaah, I. (2021) Mothers in a 'Strange Land': Migrant Women Farmers' Reproductive Health in the Brong-Ahafo Region of Ghana. *Journal of Health Care for the Poor and Underserved*, 32(2), 910–930.

Baruah, B. (2009) Monitoring progress towards gender-equitable poverty alleviation: The tools of the trade. *Progress in Development Studies*, 9(3), 171–186.

Brown, S., Roberts, J., & Taylor, K. (2010) Reservation wages, labour market participation and health. *Journal of the Royal Statistical Society: Series A (Statistics in Society)*, 173(3), 501–529.

Buettner, T., & Muenz, R. (2020) *Migration Projections: The Economic Case*. KNOMAD paper (37). https://www.knomad.org/sites/default/files/2020-02/KNOMAD_Paper_MigrationProjectionsTheEconomicCase%20_TBuettner_RMunez.pdf

Dunn, J. R., & Dyck, I. (2000) Social determinants of health in Canada's immigrant population: Results from the National Population Health Survey. *Social Science & Medicine* (1982), 51(11), 1573–1593. https://doi.org/10.1016/S0277-9536(00)00053-8

Ganle, J. K., Parker, M., Fitzpatrick, R., & Otupiri, E. (2014) Inequities in accessibility to and utilisation of maternal health services in Ghana after user-fee exemption: A descriptive study. *International Journal for Equity in Health*, 13(1). https://doi.org/10.1186/s12939-014-0089-z

Ghana Statistical Service. (2013) *Population and Housing Census, National Analytical Report*.

Heredia-Pi, I., Servan-Mori, E., Darney, B. G., Reyes-Morales, H., & Lozano, R. (2016) Measuring the adequacy of antenatal health care: A national cross-sectional study in Mexico. *Bulletin of the World Health Organization*, 94(6), 452–461. https://doi.org/10.2471/BLT.15.168302

Hosmer, Jr, D. W., Lemeshow, S., & Sturdivant, R. X. (2013) *Applied Logistic Regression* (Vol. 398). New York, NY: Wiley & Sons.

Konadu-Agyemang, K. (2000) The best of times and the worst of times: Structural adjustment programs and uneven development in Africa: The case of Ghana. *The Professional Geographer*, 52(3), 469–483.

Kuuire, V. Z., Kangmennaang, J., Atuoye, K. N., Antabe, R., Boamah, S. A., Vercillo, S., Amoyaw, J. A., & Luginaah, I. (2017) Timing and utilisation of antenatal care service in Nigeria and Malawi. *Global Public Health*, 12(6). https://doi.org/10.1080/17441692.2017.1316413

Manyeh, A. K., Amu, A., Williams, J., & Gyapong, M. (2020) Factors associated with the timing of antenatal clinic attendance among first-time mothers in rural southern Ghana. *BMC Pregnancy and Childbirth*, 20(1), 1–7. https://doi.org/10.1186/s12884-020-2738-0

Moller, A. B., Petzold, M., Chou, D., & Say, L. (2017) Early antenatal care visit: A systematic analysis of regional and global levels and trends of coverage from 1990 to 2013. *The Lancet Global Health*, 5(10), e977–e983. https://doi.org/10.1016/S2214-109X(17)30325-X

Ochako, R., Askew, I., Okal, J., Oucho, J., & Temmerman, M. (2016) Modern contraceptive use among migrant and non-migrant women in Kenya. *Reproductive Health*, 13(1), 1–8. https://doi.org/10.1186/s12978-016-0183-3

Owusu, L., & Yeboah, T. (2018) Living conditions and social determinants of healthcare inequities affecting female migrants in Ghana. *GeoJournal*, 83(5), 1005–1017. https://doi.org/10.1007/s10708-017-9817-4

Pell, C., Meñaca, A., Were, F., Afrah, N. A., Chatio, S., Manda-Taylor, L., Hamel, M. J., Hodgson, A., Tagbor, H., Kalilani, L., Ouma, P., & Pool, R. (2013) Factors affecting antenatal care attendance: Results from qualitative studies in Ghana, Kenya and Malawi. *PLoS One*, 8(1). https://doi.org/10.1371/journal.pone.0053747

Razum, O., Breckenkamp, J., Borde, T., David, M., & Bozorgmehr, K. (2018) Early antenatal care visit as indicator for health equity monitoring. *The Lancet Global Health*, 6(1), e35. https://doi.org/10.1016/S2214-109X(17)30465-5

Rishworth, A., Bisung, E., & Luginaah, I. (2016a) "It's like a disease": Women's perceptions of caesarean sections in Ghana's Upper West Region. *Women and Birth*, 29(6), e119–e125. https://doi.org/10.1016/j.wombi.2016.05.004

Rishworth, A., Dixon, J., Luginaah, I., Mkandawire, P., & Prince, C. T. (2016b) "I was on the way to the hospital but delivered in the bush": Maternal health in Ghana's Upper West Region in the context of a traditional birth attendants' ban. *Social Science & Medicine*, 148, 8–17.

Semali, I. A., Leyna, G. H., Mmbaga, E. J., & Tengia-Kessy, A. (2015) Social capital as a determinant of pregnant mother's place of delivery: Experience from Kongwa district in central Tanzania. *PLoS One*, 10(10), 1–10. https://doi.org/10.1371/journal.pone.0138887

Simkhada, B., Van Teijlingen, E. R., Porter, M., & Simkhada, P. (2008) Factors affecting the utilization of antenatal care in developing countries: Systematic review of the literature. *Journal of Advanced Nursing*, 61(3), 244–260. https://doi.org/10.1111/j.1365-2648.2007.04532.x

Songsore, J. (1979) Structural crisis, dependent capitalist development and regional inequality in Ghana. *Institute of Social Studies Occasional Papers*, 71.

Songsore, J., & Denkabe, A. (1995) *Challenging Rural Poverty in Northern Ghana: The Case of the Upper-West Region*. Trondheim: Norwegian University of Science and Technology.

Statistical Service Accra, G. (2018) *Ghana Maternal Health Survey 2017*.

Story, W. T. (2014) Social capital and the utilization of maternal and child health services in India: A multilevel analysis. *Health and Place*, 28, 73–84. https://doi.org/10.1016/j.healthplace.2014.03.011

Tekelab, T., Chojenta, C., Smith, R., & Loxton, D. (2019) The impact of antenatal care on neonatal mortality in sub-Saharan Africa: A systematic review and meta-analysis. *PLoS One*, 14(9), 1–15. https://doi.org/10.1371/journal.pone.0222566

Thangaratinam, S., Rogozińska, E., Jolly, K., Glinkowski, S., Roseboom, T., Tomlinson, J. W., Kunz, R., Mol, B. W., Coomarasamy, A., & Khan, K. S. (2012) Effects of interventions in pregnancy on maternal weight and obstetric outcomes: Meta-analysis of randomised evidence. *BMJ (Online)*, 344(7858), 1–15. https://doi.org/10.1136/bmj.e2088

UNICEF. (2020) *Under-Five Mortality*. https://data.unicef.org/topic/child-survival/under-five-mortality/, accessed 30 November 2020.

UN-SDGS. (2015) *Targets and Indicators*. https://sdgs.un.org/goals/goal3, accessed 22 November 2020.

Were, E., Nyaberi, Z., & Buziba, N. (2011) Perceptions of risk and barriers to cervical cancer screening at Moi Teaching and Referral Hospital (MTRH), Eldoret, Kenya. *African Health Sciences*, 11(1), 58–64.

WHO, UNICEF, UNFPA, World Bank, & UN. (2019) *Trends in Maternal Mortality 2000 to 2017: Estimates by WHO, UNICEF, UNFPA, World Bank Group and the United Nations Population Division*. Geneva: World Health Organization.

World Bank. (2020) *Mortality Rate, Neonatal (Per 1,000 Live Births): Sub-Saharan Africa*. https://data.worldbank.org/indicator/SH.DYN.NMRT?locations=ZG, accessed 30 November 2020.

World Health Organization. (2016) *WHO Recommendations on Antenatal Care for a Positive Pregnancy Experience*. Geneva: World Health Organization. https://apps.who.int/iris/bitstream/handle/10665/250796/9789241549912-eng.pdf.

Zhao, Q., Huang, Z. J., Yang, S., Pan, J., Smith, B., & Xu, B. (2012) The utilization of antenatal care among rural-to-urban migrant women in Shanghai: A hospital-based cross-sectional study. *BMC Public Health*, 12(1), 1–10. https://doi.org/10.1186/1471-2458-12-1012

12 Longitudinal analysis of progress in women's empowerment and maternal mortality outcomes

Evidence from Sub-Saharan Africa

Joseph Kangmennaang, Meshack Achore, Gurvir Kalsi, George A. Atiim, and Elijah Bisung

Introduction

Maternal mortality, defined as the death of a woman within 42 days of childbirth, remains a critical public health challenge globally (Ray et al., 2018). The causes of maternal mortality include a complex web of cultural, social, and political factors, as well as individual characteristics, that influence the likelihood of a woman dying during or shortly after childbirth (Choe et al., 2017). For example, a study conducted by Sajedinejad et al. (2015) showed that multifaceted factors, such as education, governance, employment and labor structure, knowledge of health service, insurance status, gender norms, health believes, and religion, are significantly associated with maternal mortality (Sajedinejad et al., 2015).

Disparities in maternal health outcomes have been a dominant feature of global health inequalities. For example, the Maternal Mortality Rate (MMR) in low- and middle-income countries (LMICs) was around 239 per 100,000 live births, compared to 12 per 100,000 live births in high-income countries in 2015 (Lan and Tavrow, 2017). Regrettably, more than half of these maternal deaths in low-income countries were recorded in SSA (Lan and Tavrow, 2017). Despite a number of global initiatives and efforts over the past decades, including successes drawn under the implementation of safe motherhood initiative (SMI) and the Millennium development goals (MDGs), maternal mortality remain a significant public health challenge in SSA, especially in rural and less accessible areas (Scott et al., 2017; Kuuire et al. 2017). For example, in 2017, Nigeria recorded the second highest number of maternal mortality deaths (40,000 maternal deaths) in the world. The high number of maternal deaths overshadowed improvements in maternal health services from 1990 to 2013 (Kalipeni et al., 2017). Further, between 1990 and 2013, Nigeria experienced an increase in the prevalence of modern contraceptive use, from 3.8% to 11.1%, and skilled birth attendance, from 30.8% to 40%, while simultaneously experiencing a decline in adolescent birth rate, from 23.5% to 17.1% (Kalipeni et al., 2017). Although these positive health outcomes presumably positively correlate with an overall decline in maternal mortality, the persistently high rates of maternal mortality in SSA only point to

DOI: 10.4324/9780367743918-15

hidden health disparities and inequities that create a foundation for poor maternal health outcomes (Kalipeni et al., 2017).

Research and practice have shown that addressing gender inequities is a crucial strategy for improving the social determinants of health that influence maternal mortality and morbidity globally, especially in SSA (Jennings et al., 2014; Adgoy, 2018). Goal 3 of the MDGs aimed to achieve gender equality and empower women to redress disadvantages through policies and programs that develop women's capabilities, improve their access to economic and political opportunities, and guarantee their safety. However, the MDG3 has been criticized for focusing on decreasing gender disparities with little emphasis on abolishing gender inequality, leaving substantive asymmetries unaddressed (Fehling et al., 2013). Thus, as countries begin to benchmark progress toward the Sustainable Development Goals (SDGs) targets, the focus must shift toward interventions that assure long-term sustainable improvements in women's health by considering gender inequities in a broader framework of economic and social progress.

In the past few years, many changes have ensued in women's lives that may be decreasing maternal mortality through women's empowerment (Lan and Tavrow, 2017). The literature put forward a variety of definitions of empowerment consisting of control over decision-making (Kabeer, 1999; Gollub, 2000); ability (agency) to make strategic choices and to regulate resources needed to attain a desired result or outcome (Alsop and Heinsohn, 2005; Kabeer, 1999); and having the power to attain desired outcomes (Grown et al., 2005; Laverack, 2006). Empowerment, according to Sen, can also be seen as an ability built on education and skills development to advocate for improved quality of life (Sen, 1990). One constituent of a better quality of life is the ability of a woman to make decisions about her body, thereby boosting positive health outcomes (Gollub, 2000).

Associations between higher levels of women's empowerment and health outcomes have been established in a burgeoning empowerment literature. For example, many investigators (Beegle et al., 2001; Hindin, 2000; Tuladhar et al., 2013) have reported associations between higher levels of empowerment and positive health outcomes. A greater degree of women's autonomy and gender equity are also reported to play an imperative role in shaping women's ability to manage fertility (Bloom et al., 2001; Sano et al., 2018), oversee the health and development of children (Shroff et al., 2009), improve nutrition and contraception use (Schuler and Hashemi, 2014), increase immunization (Thorpe et al., 2015), lower child mortality rates (Patel et al., 2006; Varkey et al., 2010), increase access to water and sanitation services (Dickin et al., 2021), and increase maternal health service utilization (Ahmed et al., 2010). Factors such as limited work opportunities, and low levels of economic, political and social participation (or lack thereof) contribute to women's disempowerment (Kantor, 2003). Consequently, the lack of empowerment results in poor maternal health, disparities in allocation of household resources, and poor medical care (Moonzwe Davis et al., 2014). According to Patel et al. (2006), gender disadvantage is one of the main determining factors of the poor health status experienced by most women globally.

In addition, women's access to and control over economic resources and participation in economic markets can reduce gender inequality and create a trajectory that leads to gender-inclusive economic growth, which results in the reduction of preventable maternal mortality (Jennings et al., 2014). However, previous studies conducted on the effectiveness of economic empowerment programs on maternal health have demonstrated conflicting perspectives. While the majority of studies (Ahmed et al., 2010; Schulz and Nakamoto, 2013; Jennings et al., 2014) report positive maternal health outcomes, others presented opposite results. For example, Rocca et al. (2009) examined the relationship between domestic violence and empowerment and found that women in south India who participated more actively in social groups and vocational training or employment opportunities, were more likely to experience domestic violence (Rocca et al., 2009). This violence, in turn, impacts negatively on the health of women (Tuladhar et al., 2013). Such gender empowerment programs failed because they did not address the more significant issue of gender norms that preserve inequalities between genders (Gupta et al., 2013). For instance, microcredit programs, which incorporate the provision of financial services for the poor (those earning less than $2/day), do not consider changing the gendered views held by male partners and, in turn, can potentially increase the rate of intimate partner violence (IPV), especially when women are financially empowered to challenge socio-familial gender norms (Van Rooyen et al., 2012; Gupta et al., 2013).

The combination of economic empowerment and health interventions has proven to be successful in reducing the structural determinants of maternal health outcomes, such as IPV (Gupta et al., 2013). The IMAGE study, which explored the incremental effects of combining health interventions (i.e., HIV training programs) and microfinance initiatives in rural South Africa, was the first trial to evaluate the impact of a combined micro-credit and participatory gender training intervention on violence (Gupta et al., 2013). The study demonstrated a 55% reduction in levels of physical and sexual intimate partner violence, reductions in levels of household poverty, and improved HIV communication (Gupta et al., 2013). Correspondingly, an integrated child nutrition program in the rural north of Vietnam, which empowers women to share information with their partners while learning vocational, problem-solving, and child care skills in women's supportive environments, improved children's food intake and reduced severe malnutrition, thereby improving both maternal and child health (Schroeder et al., 2002). This chapter aims to contribute to the bourgeoning literature on women empowerment and health by examining the ecological relationships between women's empowerment and maternal mortality among 45 SSA countries. The operationalization of empowerment is based on labor force participation and gender equality.

Method

Data

The data consist a total of 57 indicators for 54 countries in Africa, spanning from 2000 to 2015. These indicators were collected by development organizations and government institutions and were subsequently compiled by the Mo Ibrahim Foundation into a comprehensive dataset. Table 12.1 depicts the description of the variables used in the analysis. We examined whether initial levels of gender equality and changes in gender equality were associated with absence of maternal mortality ratios (AMMs). AMMs were obtained from the Maternal Mortality

Table 12.1 Descriptive Table

Indicator	*Description [range]*	*Range*
Outcome variable		
Absence of Maternal Mortality	AMM represents the risk associated with each pregnancy	[0–100] p
Gender equality	This indicator captures the extent to which the government promotes equal access for men and women to human capital development opportunities and productive and economic resources, and provides equal status and protection under the law.	Continuous [0–100]p
Labor force participation	Proportion of women economically active	[0–100]p
Gender balance in education	This indicator estimates progress toward gender parity in primary and lower secondary school enrollment.	[0-∞]p
Political participation	This indicator captures the extent to which women are represented in the legislative and executive.	
Access to improved water	The proportion of the total population that is served with an improved drinking water source, which is one that, by the nature of its construction, adequately protects the source from outside contamination, in particular with fecal matter.	[0–100]p
Access to improved sanitation	The proportion of the total population that is served with an improved sanitation facility, which is one that hygienically separates human excreta from human contact	[0–100]p
Public health campaigns	Assesses the extent to which citizens can find within easy reach abundant information about common illnesses, prevention, and treatment; the information is presented in a way that is easy to grasp for citizens with less education	[0–100]
Infrastructure	Infrastructure is one of the four sub-categories that are used to calculate the Sustainable Economic Opportunity category score. It consists of five indicators from seven data sources.	
Education	Education is one of the three sub-categories that are used to calculate the Human Development category score. It consists of eight indicators from five data sources.	
GDP	Gross national income per capita converted to international dollars ($International) using purchasing power parity rates and log transformed	

Estimation Inter-Agency Group. Our analysis included 45 countries (n = 720 country years) and excluded 9 countries due to missing data on maternal mortality and gender equality for most years. In our analysis, absence of maternal mortality per 100,000 was modeled as the outcome measure, with gender inequality, interactive effects of gender and time, and gender balance in education as predictors. We also controlled for theoretical relevant variables, such as Gross National Income per capita [log transformed], access to water and sanitation, public health campaigns, infrastructure, and education.

Analysis

We estimated two-level growth curve models to document the association between AMMR and gender equality over time. This strategy allowed us to estimate differences in initial scores and rate of growth in gender equality across the distinct waves (i.e., slopes) (Han, 2008). More specifically, we examined the effect of gender equality, and changes in gender equality on the absence of maternal mortality over a 16-year period. The level-1 equation in our growth curve models describe within-individual changes in the absence of maternal mortality over time (t) and can be represented as follows: $yt = \alpha i + \beta it + \mathcal{E}$.

Country's absence of maternal mortality (y_t) is characterized by a unique intercept (α_i) and a time-dependent slope (βi_t). Time (t) signified a series of dichotomous variables indicating the year when the data assessments were conducted to capture the non-linearity in the absence of maternal mortality. For interpretive advantage, we centered gender equality around the standard deviation; [<2SD, between 2SD and 3SD and >3SD]. Both partially adjusted (controlling for gender equality) and fully adjusted models (controlling for all covariates) were fit. The main effect models examined a linear shift in AMM (main effect models). Interaction models included an interaction term for gender equality and time, and labor force participation and time, to determine if growth trajectories varied by the levels of the absence of maternal mortality over time. The multivariable model used multi-level mixed effects linear regression, clustered at the country level. The mixed effects model, with a random intercept clustering at the country level, was chosen to account for the violation of model independence from repeated country-year measures.

Results

Descriptive results

Table 12.2 provides the descriptive results of our analysis. The total absence of maternal mortality (AMM) analysis included 45 countries that had data for all the 16 years (n = 720 country years). Figures 12.1 and 12.2 show a plot of AMM over the years, sorted by country and the average AMM over log GNI per capita. The mean absence of maternal mortality over the 16 years is 79.75 (SD = 13.31) with the range of 0–100. Mean gender equality is 58.39 (SD = 19.22) with a range of 22–100. About 24% of countries had gender equality levels less than 2 standard

Table 12.2 descriptive statistics

Variable	*Mean [range]*
Absence of maternal mortality (AMM)	Mean = 79.80, SD = 13.45 [0–100]
Gender equality	Mean = 58.44, SD = 19.22 [22.5–100]
Log of GNI per capita	Mean = 4041.01, SD = 4553.97 [532.5–22908]
Gender equality centered around the standard deviation	
$<2\sigma$	176 (24.44)
$\geq 2\sigma < 3\sigma$	220 (30.56)
$\geq 3\sigma$	324 (45.00)
Labor force participation	Mean = 62.25, SD = 26.27 [0–100]
Gender balance in education	Mean = 65.20, SD = 19.74 [6.1–100]
Political participation	Mean = 35.10, SD = 15.69 [3.2–93.4]
Time centered (ref: 2000–2003)	180 (25.00)
2004–2007	180 (25.00)
2008–2011	180 (25.00)
2012–2015	180 (25.00)
Access to water	Mean = 60.89, SD = 21.36 [7.1–100]
Sanitation	Mean = 53.53, SD = 23.43 [3–98]
Public health campaigns	Mean = 69.40, SD = 25.10 [0–100]
Infrastructure	Mean = 34.72, SD = 15.77 [5.7–92.3]
Education	Mean = 46.08, SD = 15.15 [9.6–85.3]
Observations	720
Number of countries	45

deviations below the mean, and about 44% had gender equality ratios above 3 standard deviations. The mean labor force participation rate for women was 62.77 [SD = 26.22; range: 0–100].

Multi-variate results

Table 12.3 presents the determinants of change in the absence of maternal mortality among the 45 countries over time. The findings show significant differences between AMM, initial gender equality status, change in gender equality status, as well as centered gender equality around the standard deviation. For instance, on average, there was about 0.08-point linear trend between AMM and gender equality. The statistically significant coefficient of –0.02 indicates that countries' slope differs on the absence of maternal mortality depending on its gender equality. Also, countries with gender equality between 2SD and 3SD achieved 1.16 points times higher on absence of maternal mortality than for countries below 2SD of gender equality. The results also indicate that initial labor force participation is associated with the absence of maternal mortality. Further, an increase in initial gender balance in education was positively associated with the absence of maternal mortality; however, changes in gender balance in education over time were negatively associated with the absence of maternal mortality.

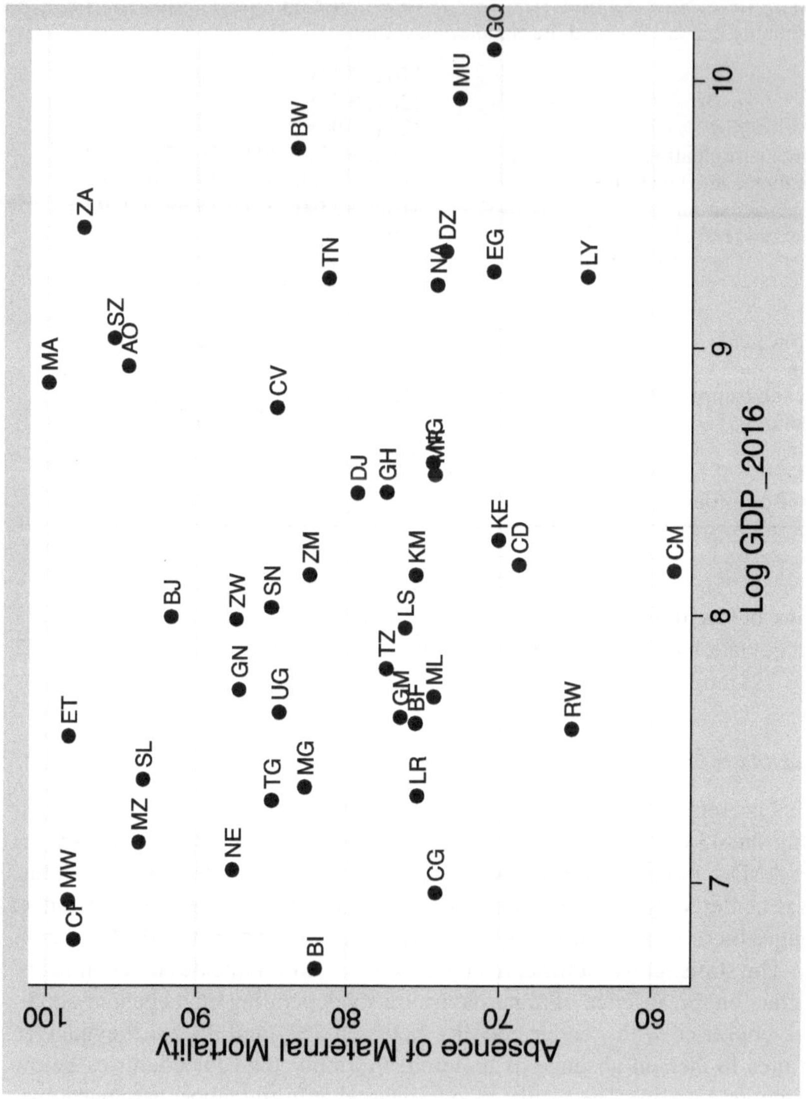

Figure 12.1 Average levels of Absence of Maternal Mortality and 2016 GDP per Capita

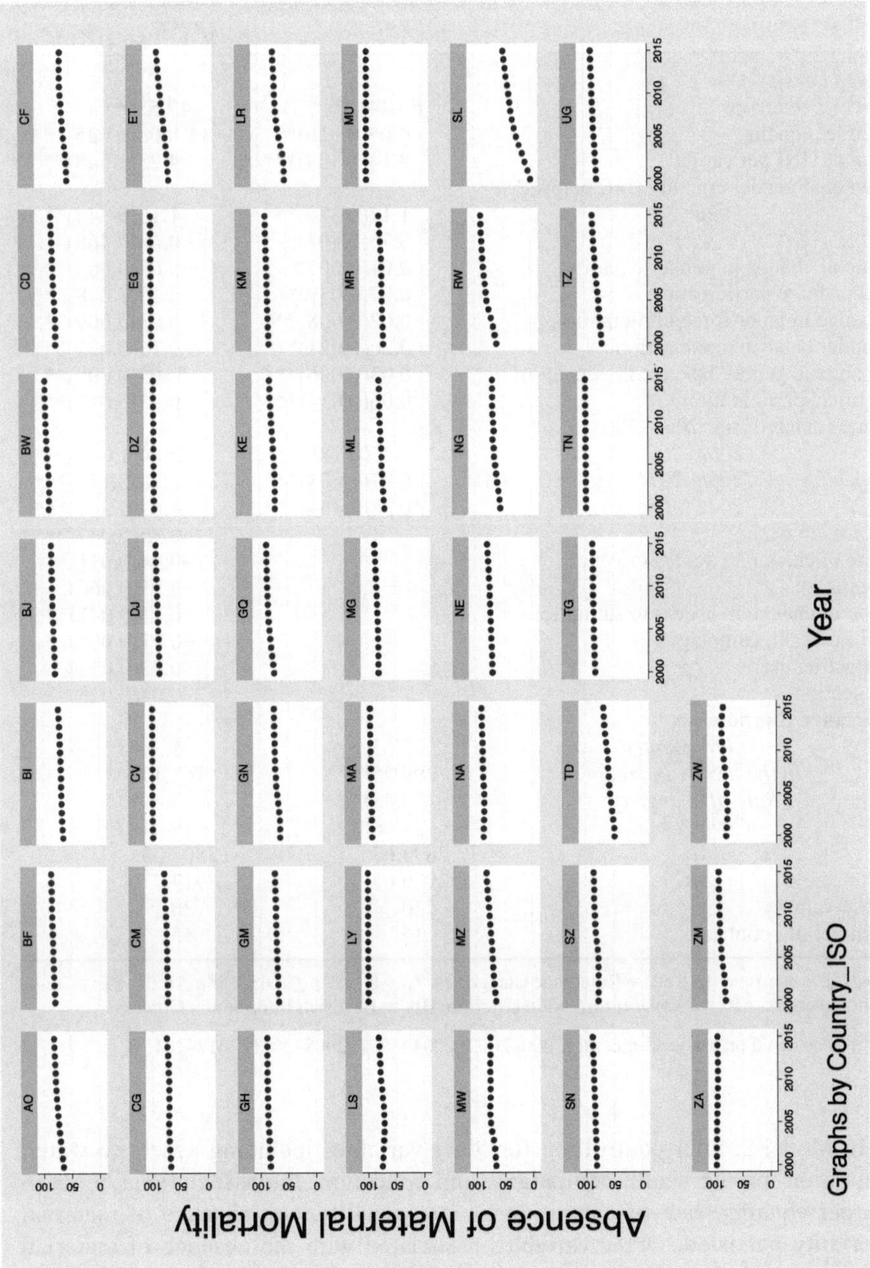

Figure 12.2 Trends in Absence of Maternal Mortality over 16 years

Table 12.3 Determinants of absence of maternal mortality among 45 countries over 16years period

Variables	*Model 1*	*Model 2 (β (SE)*
	(β (SE))	*(β (SE))*
Fixed effects		
Level-1 intercept	66.44***	47.52***
Gender equality	0.05(0.026)**	0.08(0.025)***
Log of GNI per capita	8.14(1.309)***	4.23(1.548)***
Centered gender equality (ref: below 2σ)		
≥2σ<3σ	1.17(0.370)***	1.16(0.344)***
≥3σ	0.75(0.494)	0.64(0.462)
Rate of change in gender equality	−0.01(0.007)*	−0.02(0.008)***
Labor force participation	−0.07(0.039)*	−0.10(0.038)***
Change in labor force participation	0.02(0.008)***	0.03(0.009)***
Gender balance in education	0.20(0.021)***	0.16(0.022)***
Change in gender balance in education	−0.03(0.006)***	−0.03(0.007)***
Political participation	0.06(0.014)***	0.03(0.013)***
Time centered (ref: 2000–2003)		
2004–2007	3.45(0.788)***	2.84(1.046)***
2008–2011	6.97(1.608)***	5.78(2.145)***
2012–2016	10.28(2.464)***	8.67(3.304)***
Access to water		0.35(0.0478)***
Rate of change in access to water		−0.04(0.011)***
Sanitation		−0.01(0.060)
Rate of change in access to sanitation		0.04(0.011)***
Public health campaigns		−0.01(0.005)
Infrastructure		0.02(0.022)
Education		−0.01(0.028)
Variance components		
Within country	1.65***	1.39***
In the level 1 intercept	209.25***	163.17***
Rate of change of	3.43***	3.69***
Covariance	−21.06***	−19.61***
AIC	2893.63	2803.23
BIC	2976.05	2917.71
Observations	720	720
Number of countries	45	45

Notes: β = coefficient; (Ref.) = Reference Categories; *p ≤ 0.1, **p ≤ 0.05, ***p ≤ 0.01; SE = standard errors, AIC = Akaike Information criterion, BIC = Bayesian Information Criterion

Time is centered on a 4-year interval (2000–2003; 2004–2007; 2008–2011; 2012–2015)

In Model 2, after controlling for other variables including access to water, sanitation, public health campaigns, and education, the associations between gender equality, rate of change in gender equality, and absence of maternal mortality persisted. Other variables associated with the absence of maternal mortality include access to water, rate of change in access to water, and rate of change in sanitation. The statistically significant variance components for the intercept in Model 1 ($\sigma_i^2 = 209.25$, p = 0.01) and linear change ($\sigma_1^2 = 3.43$, p = 0.001) indicate that substantial variation in these parameters is yet to be explained. In the final model (Model 2), even though the variance components for the intercept and linear change remained significant, their effects were attenuated.

Discussion

The aim of this study was to contribute to the literature on the links between women's empowerment and the absence of maternal mortality among vulnerable populations in low- to middle-income countries. In so doing, the study examined the relationships between different indicators of women's empowerment, represented by gender equality, labor force participation, gender balance in education, and political participation, with maternal mortality over a 15-year period. To this end, we used growth curve models to compare trajectories of the absence of maternal mortality in countries based on four empowerment scores, and to ascertain the role economic and social indicators play in explaining these trajectories. Even though there is an expansive research exploring the impact of gender disparities and women's empowerment on maternal health, they are predominantly country-specific and cross sectional. This tends to limit our understanding of what happens across time and how changes in gender equality and women's empowerment maybe impacting risk of maternal mortality over time. This study is one of the first to provide multi-country longitudinal evidence from the developing world, deepening our understanding of the initial and long-term effects of four key empowerment indicators on maternal mortality in 45 countries across Africa. Across the 45 countries, we observed no unique trends in the absence of maternal mortality over the 15-year period (2000 to 2015) [see Figure 12.2]. We observed marginal changes in AMM across different countries. For instance, countries with the highest reductions in absence of maternal mortality include Sierra Leone, Tanzania, Chad, Eritrea, and Liberia. The percentage reductions were very minimal among countries such as Angola and Ghana, and remained stationery in Tunisia, Togo, Sao Tome, Libya, and Mozambique. The countries that achieved greater reduction in the absence of maternal mortality are relatively less developed countries with low initial AMM values and mostly gone through conflict in the past decade (Ziegler et al., 2020).

The results show that gender equality in access to human capital development opportunities, the proportion of women who are economically active, gender parity in education, and involvement in policy-making may lead to more rapid maternal mortality decline. We found that initial levels of gender equality were associated with absence of maternal mortality and, as gender equality improves, AMM reduces. Similarly, we found that improvement in women's educational attainments and labor force participation over time was associated with lower risk of maternal mortality. These findings are consistent with recent studies that revealed that gender equality, education, and empowerment are associated with an increase in health care utilization and a reduction in maternal mortality (Choe et al., 2017; Lan & Tavrow, 2017; Souza et al., 2013; Van Der Kooi, 2013; Sano et al., 2018). For example, using a time-series analysis of maternal mortality in African countries, Van Der Kooi et al. (2013) reported that improvements in female primary school enrollment rate and socio-economic status are associated with a decline MMR over time. Even though to a large extent, the reductions in maternal deaths over the years can be attributed to global and national commitment, and investments made during the era of the MDGs (Kalipeni et al., 2017), our longitudinal analyses reveal that bridging gender gaps is one of the significant pathways toward meeting SDG 3.1 at the ***global*** and national levels and reducing maternal health inequalities in developing countries. While a suite of interventions including biomedical and community-based approaches exist (Shroff et al., 2009, Schuler and Hashemi, 2014, Thorpe et al., 2016, Patel et al., 2006; Varkey et al., 2010), we believe a greater focus on the social arrangements of power, work-life balance, and empowerment of women will sustain current maternal health gains. As our results show, there significant relationships between various indicators of gender equality (e.g., labor force participation, gender parity in education, political participation) and MMRs, even after controlling for the effects of national income, access to water, sanitation, public health campaigns, and education.

Our findings suggests a significant positive relationship between a country's per capita income and the absence of maternal mortality. This finding is similar to those of Choe et al. (2017), who found an inverse correlation between MMR and gross national per capita income. Contrary to our findings, Ensor et al. (2010) in investigating the associations between GDP per capita and changes in maternal and infant outcomes, found non-significant association between an increase in national per capita income and maternal and infant mortality (Ensor et al., 2010). Though this conflicting finding might be due to the differences in the country-years and measures (GDP versus GNP per capita), the relationship between economic progress and maternal mortality warrants further investigation. Our findings also suggest that an increase in public health campaigns has no significant effect on maternal mortality. A possible reason for this might be the design of the public health campaigns. Mass media for example is critical in disseminating public health information and improving health knowledge (Kansanga et al., 2018). However, most mass media public health campaigns do not sufficiently engage affected population and are mostly externally determined. Consequently, campaigns and education on maternal mortality prevention prove ineffective or fail.

This analysis is not without limitations. First, due to the ecological nature of the data and our analysis, we were unable to make any inference at the individual level. Second, the data were obtained from publicly available sources and we cannot guarantee its accuracy. However, we do respect and appreciate the efforts made by institutions like the Mo Ibrahim Foundation, in generating data on social progress in SSA. Third, we could not include all factors that are deemed to affect maternal health, as indicated in previous literature. Fourth, due to missing data, we could not include data on all countries in SSA, but were able to include data from 85% of these countries, as well as the theoretically relevant covariates. Due to these limitations, the interpretation of the results should be done with caution.

In conclusion, our results have implications for the Global Health 2035 report, and the ambitious Sustainable Development Goals (United Nations SDGS, 2015; Jamison et al., 2013). The MDG targets were mostly not met in many low-income countries and the SDG targets are more ambitious, so we clearly need corresponding ambitious policy innovation. In fact, this chapter shows that SDG 3.1 for reducing maternal mortality is complementary to SDG 5.5 for raising the share of women in parliament. Our findings suggest that a reduction in maternal mortality can be achieved by curtailing gender discrimination. Thus, effectively reducing the gender gap in education, economic welfare and policy-making. Inasmuch as clinical and biomedical interventions, such as improvements in obstetric care and access to a skilled birth attendant, are important, our findings show that an improvement in women's economic empowerment and education are an important preventive measure in reducing maternal mortality, especially in Sub-Sahara Africa.

References

Adgoy, E.T. (2018) Key Social Determinants of Maternal Health among African Countries: A Documentary Review. *MOJ Public Health*, 7(3).

Ahmed, S., Creanga, A.A., Gillespie, D.G. and Tsui, A.O. (2010) Economic Status, Education and Empowerment: Implications for Maternal Health Service Utilization in Developing Countries. *PLoS One*, 5(6).

Alsop, R. and Heinsohn, N. (2005) *Measuring Empowerment in Practice: Structuring Analysis and Framing Indicators* [online]. World Bank Publications. Available at: http://siteresources.worldbank.org/INTEMPOWERMENT/Resources/41307_wps3510.pdf, accessed 19 Mar. 2021.

Beegle, K., Frankenberg, E. and Thomas, D. (2001) Bargaining Power within Couples and Use of Prenatal and Delivery Care in Indonesia. *Studies in Family Planning*, 32, 130–146.

Bloom, S.S., Wypij, D. and das Gupta, M. (2001) Dimensions of Women's Autonomy and the Influence on Maternal Health Care Utilization in a North Indian City. *Demography*, 38, 67.

Choe, S.-A., Cho, S. and Kim, H. (2017) Gender Gap Matters in Maternal Mortality in Low and Lower-Middle-Income Countries: A Study of the Global Gender Gap Index. *Global Public Health*, 12, 1065–1076.

Dickin, S., Bisung, E., Nansi, J. and Charles, K. (2021) Empowerment in Water, Sanitation and Hygiene Index. *World Development*, 137, 105158.

Ensor, T., Cooper, S., Davidson, L., Fitzmaurice, A. and Graham, W.J. (2010) The Impact of Economic Recession on Maternal and Infant Mortality: Lessons from History. *BMC Public Health*, 10.

Fehling, M., Nelson, B.D. and Venkatapuram, S. (2013) Limitations of the Millennium Development Goals: A Literature Review. *Global Public Health*, 8, 1109–1122.

Gollub, E. L. (2000) The Female Condom: Tool for Women's Empowerment. *American Journal of Public Health*, 90, 1377–1381.

Grown, C., Gupta, G. R. and Pande, R. (2005) Taking Action to Improve Women's Health through Gender Equality and Women's Empowerment. *Lancet*, 365, 541–543.

Gupta, J., Falb, K.L., Lehmann, H., Kpebo, D., Xuan, Z., Hossain, M., Zimmerman, C., Watts, C. and Annan, J. (2013) Gender Norms and Economic Empowerment Intervention to Reduce Intimate Partner Violence against Women in Rural Côte d'Ivoire: A Randomized Controlled Pilot Study. *BMC International Health and Human Rights*. Available at: www.ncbi.nlm.nih.gov/pmc/articles/PMC3816202/, accessed 19 Jul. 2019.

Han, W.J. (2008) The Academic Trajectories of Children of Immigrants and Their School Environments. *Developmental Psychology*, 44(6), 1572.

Hindin, M.J. (2000) Women's Power and Anthropometric Status in Zimbabwe. *Social Science & Medicine*, 51, 1517–1528.

Jamison, D.T., Summers, L.H., Alleyne, G., Arrow, K.J., Berkley, S., Binagwaho, A., Bustreo, F., Evans, D., Feachem, R.G.A., Frenk, J., Ghosh, G., Goldie, S.J., Guo, Y., Gupta, S., Horton, R., Kruk, M.E., Mahmoud, A., Mohohlo, L.K., Ncube, M. and Pablos-Mendez, A. (2013) Global Health 2035: A World Converging within a Generation. *The Lancet*, 382, 1898–1955.

Jennings, L., Na, M., Cherewick, M., Hindin, M., Mullany, B. and Ahmed, S. (2014) Women's Empowerment and Male Involvement in Antenatal Care: Analyses of Demographic and Health Surveys (DHS) in Selected African Countries. *BMC Pregnancy and Childbirth*, 14.

Kabeer, N. (1999) Resources, Agency, Achievements: Reflections on the Measurement of Women's Empowerment. *Development and Change*, 30, 435–464.

Kalipeni, E., Iwelunmor, J. and Grigsby-Toussaint, D. (2017) Maternal and Child Health in Africa for Sustainable Development Goals beyond 2015. *Global Public Health*, 12, 643–647.

Kansanga, M.M., Asumah Braimah, J., Antabe, R., Sano, Y., Kyeremeh, E. and Luginaah, I. (2018) Examining the Association between Exposure to Mass Media and Health Insurance Enrolment in Ghana. *The International Journal of Health Planning and Management*, 33, 531–540.

Kantor, P. (2003) Women's Empowerment through Home-Based Work: Evidence from India. *Development and Change*, 34, 425–445.

Kuuire, V.Z., Kangmennaang, J., Atuoye, K.N., Antabe, R., Boamah, S.A., Vercillo, S., Amoyaw, J.A. and Luginaah, I. (2017) Timing and Utilisation of Antenatal Care Service in Nigeria and Malawi. *Global Public Health*, 12, 711–727.

Lan, C.-W. and Tavrow, P. (2017) Composite Measures of Women's Empowerment and Their Association with Maternal Mortality in Low-Income Countries. *BMC Pregnancy and Childbirth*, 17.

Laverack, G. (2006) Improving Health Outcomes through Community Empowerment: A Review of the Literature. *Journal of Health, Population and Nutrition*, 24, 113–120.

Moonzwe Davis, L., Schensul, S.L., Schensul, J.J., Verma, R.K., Nastasi, B.K. and Singh, R. (2014) Women's Empowerment and Its Differential Impact on Health in Low-Income Communities in Mumbai, India. *Global Public Health*, 9, 481–494.

Patel, V., Kirkwood, B.R., Pednekar, S., Pereira, B., Barros, P., Fernandes, J., Datta, J., Pai, R., Weiss, H. and Mabey, D. (2006) Gender Disadvantage and Reproductive Health Risk Factors for Common Mental Disorders in Women. *Archives of General Psychiatry*, 63, 404.

Ray, J.G., Park, A.L., Dzakpasu, S., Dayan, N., Deb-Rinker, P., Luo, W. and Joseph, K.S. (2018) Prevalence of Severe Maternal Morbidity and Factors Associated with Maternal Mortality in Ontario, Canada. *JAMA Network Open*, 1, e184571.

Rocca, C.H., Rathod, S., Falle, T., Pande, R.P. and Krishnan, S. (2009) Challenging Assumptions about Women's Empowerment: Social and Economic Resources and Domestic Violence among Young Married Women in Urban South India. *International Journal of Epidemiology*, 38, 577–585.

Sajedinejad, S., Majdzadeh, R., Vedadhir, A., Tabatabaei, M. and Mohammad, K. (2015) Maternal Mortality: A Cross-Sectional Study in Global Health. *Globalization and Health*, 11, 4.

Sano, Y., Antabe, R., Atuoye, K.N., Braimah, J.A., Galaa, S.Z. and Luginaah, I. (2018) Married Women's Autonomy and Post-Delivery Modern Contraceptive Use in the Democratic Republic of Congo. *BMC Women's Health*, 18, 49.

Schroeder, D.G., Pachón, H., Dearden, K.A., Ha, T.T., Lang, T.T. and Marsh, D.R. (2002) An Integrated Child Nutrition Intervention Improved Growth of Younger, More Malnourished Children in Northern Viet Nam. *Food and Nutrition Bulletin*, 23, 50–58.

Schuler, S.R. and Hashemi, S.M. (2014) Credit Programs, Women's Empowerment, and Contraceptive Use in Rural Bangladesh. *Studies in Family Planning*, 25, 65–76.

Schulz, P.J. and Nakamoto, K. (2013) Health Literacy and Patient Empowerment in Health Communication: The Importance of Separating Conjoined Twins. *Patient Education and Counseling*, 90, 4–11.

Scott, S., Kendall, L., Gomez, P., Howie, S.R.C., Zaman, S.M.A., Ceesay, S., D'Alessandro, U. and Jasseh, M. (2017) Effect of Maternal Death on Child Survival in Rural West Africa: 25 Years of Prospective Surveillance Data in the Gambia. *PLoS One*, 12, e0172286.

Sen, A. (1990) Gender and Cooperative Conflicts. In I. Tinker (Ed.), *Persistent Inequalities: Women and World Development*. New York: Oxford University Press, 123–149.

Shroff, M., Griffiths, P., Adair, L., Suchindran, C. and Bentley, M. (2009) Maternal Autonomy Is Inversely Related to Child Stunting in Andhra Pradesh, India. *Maternal & Child Nutrition*, 5, 64–74.

Souza, J.P., Gülmezoglu, A.M., Vogel, J., Carroli, G., Lumbiganon, P., Qureshi, Z., Costa, M.J., Fawole, B., Mugerwa, Y., Nafiou, I., Neves, I., Wolomby-Molondo, J.-J., Bang, H.T., Cheang, K., Chuyun, K., Jayaratne, K., Jayathilaka, C.A., Mazhar, S.B., Mori, R. and Mustafa, M.L. (2013) Moving beyond Essential Interventions for Reduction of Maternal Mortality (the WHO Multicountry Survey on Maternal and Newborn Health): A Cross-Sectional Study. *The Lancet*, 381, 1747–1755.

Thorpe, S., VanderEnde, K., Peters, C., Bardin, L. and Yount, K.M. (2016) The Influence of Women's Empowerment on Child Immunization Coverage in Low, Lower-Middle, and Upper-Middle Income Countries: A Systematic Review of the Literature. *Maternal and Child Health Journal*, 20, 172–186.

Tuladhar, S. *et al.* (2013) *Women's Empowerment and Spousal Violence in Relation to Health Outcomes in Nepal, Further Analysis of the 2011 Nepal Demographic and Health Survey*. Available at: https://dhsprogram.com/pubs/pdf/FA77/FA77.pdf, accessed 17 Mar. 2019.

United Nations SDGS. (2015) *The 17 Goals* [online]. United Nations. Available at: https://sdgs.un.org/goals, accessed 18 Mar. 2021.

van der Kooi, A.L.F., Stronks, K., Thompson, C.A., DerSarkissian, M. and Arah, O.A. (2013) The Modifying Influence of Country Development on the Effect of Individual Educational Attainment on Self-Rated Health. *American Journal of Public Health*. Available at: www.ncbi.nlm.nih.gov/pmc/articles/PMC3828725/, accessed 30 Oct. 2019.

van Rooyen, C., Stewart, R. and de Wet, T. (2012) The Impact of Microfinance in Sub-Saharan Africa: A Systematic Review of the Evidence. *World Development*, 40, 2249–2262.

Varkey, P., Kureshi, S. and Lesnick, T. (2010) Empowerment of Women and Its Association with the Health of the Community. *Journal of Women's Health*, 19, 71–76.

Ziegler, B.R., Kansanga, M., Sano, Y., Kangmennaang, J., Kpienbaareh, D. and Luginaah, I. (2020) Antenatal Care Utilization in the Fragile and Conflict-Affected Context of the Democratic Republic of the Congo. *Social Science and Medicine*, 262, 113253.

13 Mental health, quality of life, and life experiences of Ghanaian women living with breast cancer

Rhonda Boateng, Beatrice Wiafe-Addai, Allison Williams, & Harry Shannon

Introduction

Projected to rise by more than 85% by 2030, cancer will claim the lives of nearly one million SSAs (Morhason-Bello et al., 2013; Sylla & Wild, 2012). In SSA, breast cancer is the second most common form of cancer and a leading cause of death among women, accounting for 16% of cancer-related deaths (Morhason-Bello et al., 2013). Despite having a lower incidence of breast cancer (41.5 versus 89.4 cases per 100,000 women per year), the fatality rate for West African women is significantly higher than that in North America (22.3 versus 12.5 cases per 100,000 women) (Sung et al., 2021).

Breast cancer is the leading cause of cancer mortality among Ghanaian women with an incidence rate of 38.3 cases per 100,000 people, and a mortality rate of 18.1/100,000 (Sung et al., 2021). Despite the high mortality rate, breast cancer remains largely curable. Yet late-stage diagnosis and treatment default are contributors to high mortality (Obrist et al., 2014; Ohene-Yeboah & Adjei, 2012). Between 60% and 85% of women are diagnosed in advanced stages of breast cancer, averaging 10 months of experiencing symptoms by the time of hospital presentation (Dedey et al., 2016; Scherber et al., 2014; Ohene-Yeboah & Adjei, 2012). Furthermore, about a quarter of Ghanaian patients' default on treatment (Clegg-Lamptey et al., 2009). Predictors of treatment default include seeking treatment with traditional healers, lack of knowledge specific to national health insurance coverage for breast cancer treatment, lack of counseling, and lack of emotional support from health professionals (Scherber et al., 2014; Clegg-Lamptey et al., 2009). As incidence continues to increase in Ghana, treatment unaffordability, geographical barriers, and lack of knowledge about breast cancer must be addressed (McKenzie et al., 2018; Zelle et al., 2012). Treatment centers are concentrated in major cities, obliging some women to travel long distances to receive medical attention. As well, treatment costs are high and the lack of breast cancer education means women do not learn to recognize the signs and symptoms of cancer (McKenzie et al., 2018; Brakohiapa et al., 2013).

Physical health and mental health interact in a bidirectional manner where the aggravation of one can deeply impact the other. Individuals with chronic illnesses, such as breast cancer, are at an increased risk of developing a mental disorder (Patel &

DOI: 10.4324/9780367743918-16

Kleinman, 2003; De Menil et al., 2012). Conversely, psychological disorders are strong predictors of the development of physical illness (Prince et al., 2007; Sweetland et al., 2014). Zainal et al. (2007) found that among 168 Malaysian women undergoing chemotherapy, 51% experienced psychological distress, and 32% had depression and anxiety. Depression and anxiety are common mental disorders affecting Ghanaian breast cancer patients (Kyei et al., 2020). Women with breast cancer experience stigma and changes to body image and social roles, both which increase risk to metal disorders and psychological distress (Distelhorst et al., 2015; Akin-Odanye et al., 2011). Psychological distress and mental disorders promote treatment default, amplify functional impairment, and negatively impact quality-of-life, disease recurrence-free intervals, survival, and prognosis (Kugbey et al., 2019; Distelhorst et al., 2015; Groenvold et al., 2007; Fann et al., 2008).

Improving the condition of women living with breast cancer can facilitate achieving a number of SDGs, including: No Poverty (SDG 1), Good Health and Well-Being (SDG 3), Gender Equality (SDG 5) and Decent Work and Economic Growth (SDG 8). Notably, there is a mandate to reduce mortality from non-communicable diseases, such as breast cancer, and promote mental health. Breast cancer peaks between 35 and 45 years in West African women (10–15 years earlier than for Western countries), largely affecting premenopausal women still in the workforce (Brakohiapa et al., 2013). As it can impede a woman's ability to work, having breast cancer can hinder equal access to economic resources (SDG 5.A) and sustainable economic growth and productivity (SDG 8.1 and SDG 8.2) (Igene, 2008). Moreover, physical, and mental illness can increase the chances of loss of employment, leading to income reduction and, in some cases, to poverty (Skeen et al., 2010). Furthermore, the children of mothers with poor mental health had increased likelihood of "poor nutrition, stunting, early cessation of breastfeeding, infectious disease and diarrheal disease" (Adewuya et al., 2008; Miranda & Patel, 2005). Hence, addressing women's health will also help with reducing child mortality.

While evidence indicates that physical and mental health issues work in a symbiotic manner to compromise quality-of-life, scant research has been done on the topic in Africa (Berard et al., 1998). In response, this study aims to assess the psychological distress, lived experiences and quality of life of Ghanaian women living with breast cancer. A thorough understanding of barriers and the lived experiences of women with breast cancer can identify potential actions that address gaps in disease management and ensure Ghana attains its goal of achieving the SDGs.

Methods

We conducted a sequential explanatory mixed-methods study involving both quantitative and qualitative components. A cross-sectional design (quantitative) was used to assess psychological distress and quality of life as primary outcomes. Qualitative findings were collected via interviews, and thematic analysis of interview data further explained the quantitative results. Inspired by Gonzaga (2013) and adapted to the Ghanaian context, the interview questions followed phenomenological inquiry and centered on the women's knowledge of breast cancer and coping mechanisms. All participants provided consent. Ethics approval was obtained from McMaster University (Canada) and Kwame Nkrumah University of Science and Technology (Ghana).

Participants

Women over 18 years old with comprehension of English or Twi were eligible participants. Women with breast cancer, recruited from the Peace & Love Hospitals (PLH), must have been diagnosed in the last five years, and have no comorbidity. The controls, recruited from PLH and breast cancer screening events, must not have been diagnosed with a major illness and were selected to match the age distribution of women with cancer. Participants were informed about the study by a research team member. Those who met the eligibility criteria were invited to participate in the study. All participants received 10 GH₵ (about $3.50 CND). Data were collected from June to August 2016.

Using a two-sided $\alpha = 0.05$ and power of 0.80 to detect a medium-sized effect between two groups, the required sample size was 64 patients and 64 controls (Cohen, 1992). See descriptive statistics listed in Table 13.1. To match the survey sample, a smaller subset of breast cancer patients was selected for an interview, based on age, marital status, and education. Recruitment of interviewees stopped upon data saturation.

Table 13.1 Descriptive Statistics for study participants

	Breast Cancer	*Controls*
	N	*N*
Region		
Greater Accra	20	14
Ashanti	21	20
Brong-Ahafo	4	1
Eastern	8	13
Western	3	12
Central	3	0
Volta	3	4
Upper Western	1	0
Northern	1	0
Education		
None	10	4
Elementary school	25	22
High school	10	17
Tertiary	19	20
Marital Status		
Single	13	17

(*Continued*)

Table 13.1 (Continued)

	Breast Cancer	*Controls*
	N	*N*
Married	40	35
Widow	11	11
Cancer Stage		
Early	11	
Late	19	
Not aware	34	
Cancer Treatment		
No treatment	7	
Chemotherapy	45	
Breast operation	31	
Radiotherapy	15	
Hormone therapy	9	

Questionnaires

Participants were asked to answer a demographic questionnaire, the World Health Organization Quality of Life BREF (WHOQoL-BREF), and the Kessler Psychological Distress scale (K-10) in either English or Twi. Participants disclosed their place of residence, education level, date of birth, marital status, and major health issues. The type of cancer treatment and cancer stage was also collected. The WHOQoL-BREF is a valid and reliable quality-of-life scale, recommended for cross-sectional studies (Bowden & Fox-Rushby, 2003; Harper & WHOQOL Group, 1996). Quality of life is defined as an "individuals' perceptions of their position in life in the context of the culture and value systems in which they live and in relation to their goals, expectations, standards and concerns" (Harper & WHOQOL Group, 1996). The measure has four domains: physical health, psychological well-being, social relationships, and environment. Scoring followed the formal WHOQoL-BREF procedure (WHO, 1996). Each domain was scored on a 20-point scale.

Self-reported quality of life (item 1) and the score for quality of life and satisfaction with health (a combination of items 1 and 2) were converted into a dichotomous variable (low and high) using a median split. For the self-reported quality-of-life score, women who scored 75 or 100 were marked as high, and those who scored 0, 25 or 50 were labeled low. The quality of life and satisfaction with health score ranged from 4 to 20 and scores below 16 were labeled low.

The K-10, a robust measure of non-specific psychological distress, was validated in a SSA population (Bougie et al., 2016). Scores under 20 were labeled as doing well, those between 20 and 24 indicated a mild mental disorder, and those between 25 and 29 designated a moderate mental disorder. Women who scored over 30 were considered to have a severe mental disorder (Andrews & Slade, 2001).

Quantitative analysis

Multiple regressions were run with the quality-of-life domains and psychological distress as dependent variables, and cancer diagnosis, age, education, type of treatment, and marital status as independent variables. Logistic regressions were performed to ascertain the relationships between cancer diagnosis, age, marital status, education, and type of treatment, with quality-of-life and satisfaction with health. To compare the K-10 and WHOQoL-BREF responses based on cancer diagnosis, independent-samples t-tests were conducted.

Age was a continuous variable and cancer diagnosis (control or breast cancer) was a dichotomous variable. Marital status was grouped into three categories: single, married (reference), and widowed. There were four categories of education: no education (reference), elementary, high school, and tertiary. Types of treatment were treated as separate variables and included: chemotherapy, breast surgery, radiotherapy, and hormone therapy. As 34 participants (53%) were not aware of their cancer stage, that variable was omitted from the analysis.

Qualitative interviews & analysis

We conducted semi-structured interviews to explore the lived experiences of women living with breast cancer. We also explored the impact of breast cancer on: everyday activities, community/family roles and responsibilities, and female identity. Interviews, averaging 20 minutes, were audio-recorded and transcribed. The researcher (RB) wrote down a priori knowledge and hypotheses in field notes and practiced bracketing (Richards & Morse, 2007; Giacomini, 2010). Critical reflexivity was performed throughout (Hay, 2010). Transcripts and observations were coded. Codes were then grouped into themes to perform within-case and cross-case descriptive and thematic analyses. Data saturation was achieved when no new themes emerged. Pseudonyms are used throughout.

Results

Quantitative findings

Quality-of-life and psychological distress

The regressions showed cancer diagnosis to be a significant correlate of psychological distress and quality-of-life domains, except for social relationships and self-reported quality of life (Tables 13.2 and 13.3). On average, breast cancer patients rated their physical health 1.95 [95% CI = .96, 2.93] points lower than the women in the control. They were less satisfied with their ability to perform their daily living activities [Mean Difference (MD) = .55, 95% CI = .18, .9]), and experienced more physical pain than the controls (MD = −1.23, 95% CI = −1.61, −. 85). In the previous 30 days, breast cancer patients and women in the control

Table 13.2 Multiple Regressions Results

Independent Variables	*Dependent variables*									
	Physical Health		*Psychological Well-being*		*Social Relationships*		*Environment*		*Psychological Distress*	
	B-value	*95% CI*	*B-value*	*95% CI*	*B-value*	*95% CI*	*B-value*	*95% CI*	*B-value*	*95% CI*
Education (no education)										
Elementary	.836	−.83, 2.5	.23	−1.63, 2.09	−1.16	−3.15, .834	−.57	−2.18, 1.03	.3	−4.21, 4.81
High school	1.11	−.74, 2.9	1.9	−.22, 3.94	.29	−1.9, 2.49	1.02	−.77, 2.8	−3.09	−8.12, 1.94
Tertiary	1.04	−.67, 2.76	3.19	1.27, 5.1	.48	−1.57, 2.52	1.88	.23, 3.52	−4.97	−9.6, −.33
Marital status (single)										
Married	.35	−.84, 1.54	.11	−1.23, 1.45	−.87	−2.25, .52	.24	−.91, 1.38	.54	−2.67, 3.75
Widowed	.43	−1.21, 2.07	−.64	−2.49, 1.2	.39	−2.33, 1.55	−.71	−2.29, .87	−.83	−5.27, 3.6
Cancer diagnosis (control)	−1.95	−2.93, −.96	−1.57	−2.67, −.47	.064	−1.08, 1.2	−1.21	−2.16, −.27	2.95	.29, 5.6
Age- 10-year increase	−.19	−.67, .29	−.02	−.56, .52	.05	−.51, .61	.02	−.45, .48	.2	−1.11, 1.5

Note: Reference categories for variables are in parentheses.

Table 13.3 Logistic Regression Results

Independent Variables	*Dependent Variables*			
	Overall Quality of Life		*Overall Quality of Life & Satisfaction with Health*	
	Odds Ratio	*95% CI*	*Odds Ratio*	*95% CI*
Education (no education)				
Elementary	1.28	.37, 4.41	.95	.26, 3.64
High school	1.58	.39, 6.46	1.54	.34, 6.37
Tertiary education	9.45	2.07, 43.19	5.55	1.35, 21.66
Marital status (single)				
Married	1.51	.57, 3.98	1.55	.37, 4.59
Widowed	.6	.16, 2.32	1.07	.18, 4.39
Cancer diagnosis (control)	.57	.25, 1.3	.32	.15, .72
Age 10-year increase	1.105	.817, 1.79	.904	.599, 1.344

Note: Reference categories for variables are in parentheses.

reported that they, on average, were unable to work for nine days and 2.5 days, respectively (MD = −6.4, 95% CI = −9.59, −3.1).

Women with cancer had a 1.57 [95% CI = 2.67, .47] lower mean score on psychological well-being domain, and a 2.95 [95% CI = .29, 5.6] higher score on psychological distress than the controls. Breast cancer patients reported more difficulty accepting their bodily appearance than women in the control group (MD = .89, [95% CI = .44, 1.33]). The women with cancer also felt negative feelings, such as anxiety and depression (MD = −.71, 95% CI = −1.06, −.36), and attributed the cause of their psychological distress to physical health problems more often than women in the control (MD = −.69, 95% CI = −1.1, −.29). Based on the K-10 score, women with breast cancer, on average, experienced moderate mental disorder.

Breast cancer patients had a 1.21 [95% CI = 2.16, .27] lower mean score on the environment domain scale than the individuals in the control group. Furthermore, women with breast cancer had fewer opportunities for leisure activities (MD = .61, 95% CI = .17, 1.05), and reported feeling they did not have enough money to meet their needs more often than those in the control (MD = .63, 95% CI = .23, 1.02).

Among the women with breast cancer, the multiple linear and logistic regressions for quality of life and psychological distress showed significant relationships with education, type of treatment, and age (Tables 13.4 and 13.5). No independent variable significantly predicted scores for physical health and environment. Education significantly predicted psychological well-being [partial $F(3,51) = 5.72$, $p = .002$]. Overall, women with higher education were more likely to report better psychological well-being. Women with high school education had a 3.39 [95% CI

Table 13.4 Results of the multiple regressions within breast cancer population

	Dependent Variables									
	Physical Health		*Psychological Well-being*		*Social Relationships*		*Environment*		*Psychological distress*	
Independent Variables	*B-value*	*95% CI*	*B-value*	*95% CI*	*B-value*	*95% CI*	*B-value*	*95% CI*	*B-value*	*95% CI*
Education (no education)										
Elementary	.83	−1.5, 3.14	.97	−1.34, 3.29	.45	−1.96, 2.86	.73	−1.39, 2.85	−.57	−5.98, 4.83
High school	1.83	−.99, 4.65	3.39	.59, 6.19	2.02	−.89, 4.94	2.0	−.57, 4.56	−3.65	−10.18, 2.88
Tertiary	.7	−1.72, 3.12	4.05	1.65, 6.45	.11	−2.39, 2.61	2.33	.13, 4.52	−4.24	−9.84, 1.37
Marital status (single)										
Married	.62	−1.34, 2.58	.44	−1.51, 2.39	−2.17	−4.2, −.14	.15	−1.63, 1.94	−2.93	−7.47, 1.62
Widow	.096	−2.81, 3.0	−1.36	−4.24, 1.53	−3.26	−6.26, −.26	−1.64	−4.28, 1.0	−.74	−7.46, 5.99
Treatment										
Breast surgery	.59	−1.03, 2.21	1.54	−.07, 3.15	.18	−1.49, 1.86	1.39	−.09, 2.86	−2.48	−6.23, 1.28
Radiotherapy	.69	−1.2, 2.57	2.29	.42, 4.16	1.42	−.52, 3.37	.7	−1.02, 2.41	4.89	.53, 9.25
Hormone therapy	−1.31	−3.6, .95	−1.72	−3.96, .52	−.47	−2.81, 1.86	.51	−1.55, 2.56	−3.38	−8.62, 1.85
Chemotherapy	.25	−1.56, 2.07	−1.46	−3.25, .34	−.86	−2.73, 1.01	−1.01	−2.66, .63	−3.04	−7.24, 1.16
Age 10-year increase	−.04	−.80, .71	.44	−.31, 1.19	.81	.03, 1.60	.02	−.67, .71	−.28	−2.04, 1.48

Note: Reference categories for variables are in parentheses.

Table 13.5 Results of the logistic regressions within breast cancer population

Independent Variables	*Overall Quality of Life*		*Overall Quality of Life & Satisfaction with Health*	
	Odds Ratio	*95% CI*	*Odds Ratio*	*95% CI*
Education				
(no education)				
Elementary	11.34	1.07, 119.84	2.2	.25, 31.97
High school	33.32	1.57, 706.6	23.5	1.03, 534.9
Tertiary	61.58	3.65, 1204.41	12.88	1.05, 158.6
Marital status				
(married)				
Single	.53	.11, 2.72	1.33	.17, 3.4
Widowed	.13	.02, 1.11	.64	.05, 4.27
Treatment				
Breast surgery	6.4	1.15, 35.56	11.26	1.92, 65.97
Radiotherapy	5.76	.78, 42.68	11.07	1.67, 73.31
Hormone therapy	1.38	.2, 11.4	1.5	.24, 9.64
Chemo	.09	.01, .64	.27	.04, 1.85
Age 10-year increase	2.24	1.040, 4.77	1.552	.808, 2.997

= .59, 6.19] higher psychological well-being score than those with no education; tertiary educated women had a 4.05 [95% CI = 1.65, 6.44] higher score. Cancer treatment was also significantly associated with improved psychological well-being [partial $F(4,51) = 3.51$, $p = .01$]. Compared to those who did not receive radiotherapy, participants who received this treatment had a 2.3 [95% CI = .42, 4.16] higher score for psychological well-being, and a 4.89 [95% CI = .53, 9.25] higher score for psychological distress.

Women who received chemotherapy were less likely [Odds Ratio (OR) = .09; 95% CI = .014, .64] to self-report their overall quality of life as high. However, women who had breast surgery were 6.4 times [95% CI = 1.2, 35.6] more likely to do so. High self-reported quality of life increased with age [OR = 2.24, 95% CI = (1.040, 4.77)], for a 10-year increase in age. Though weak, education was a significant predictor, as those with higher levels of education tended to report their quality of life as high ($\chi^2(3) = 8.37$, $p = .04$).

Women who had breast surgery were 11.26 [95% CI = 1.92, 65.97] times more likely, and women who received radiotherapy were 11.07 [95% CI = 1.7, 73.3] times more likely than those who did not receive these treatments, to rate their overall quality of life and satisfaction with health as high; however, the confidence intervals were wide due to the small sample size.

Qualitative findings

Thirteen women with breast cancer were included in the qualitative arm of the study (Table 13.6). Patients typically followed a similar four-stage journey:

Table 13.6 Demographic Table of Interviewed Participants

Pseudonym	*Age*	*Education*	*Marital Status*	*Residence*
Bridget	59	No education	Married	Ashanti
Nancy	50	Elementary school	Divorced	Eastern
Jane	57	Tertiary	Married	Central
Mavis	46	High school	Married	Ashanti
Talia	38	High school	Married	Brong-Ahafo
Lynn	58	Tertiary	Married	Eastern
Lois	52	Tertiary	Married	Ashanti
Gifty	69	Tertiary	Widowed	Greater Accra
Patricia	40	Tertiary	Married	Upper West
Mansah	41	Tertiary	Divorced	Ashanti
Ann	56	Tertiary	Married	Greater Accra
Bernice	72	Elementary	Married	Greater Accra
Mary	66	Tertiary	Married	Greater Accra

knowledge and suspicion, navigating the health system, the impact of breast cancer, and regaining confidence.

Knowledge and suspicion

Among the interviewees, breast cancer knowledge prior to diagnosis ranged from quite knowledgeable to having no knowledge. Six women knew about breast cancer through a relative with the disease and/or through awareness campaigns. Two women had little to no knowledge about the disease until diagnosis.

For most women, the first suspicion of ill health arose after feeling a breast lump and/or pain. Two women initially dismissed the symptoms until their deteriorating condition compelled them to seek medical attention. Suspicion of ill health did not arise from clinical breast screening, as none had a history of routine screenings.

Navigating the Ghanaian health system

After suspicion of ill health, the women navigated through the Ghanaian health system. Many participants did not know where to go, and found obtaining information difficult. Referrals to the hospital and access to breast cancer information were often provided by friends and family. Often, women went through multiple nodes of care before starting cancer care. Mary was transferred between three hospitals before receiving cancer treatment. Mary recounted feeling dehumanized by healthcare workers. A botched biopsy led to a foul-smelling and leaking wound that gravely affected her physically and psychologically.

> They were using me for learning. They would play with me
>
> – Mary

Being diagnosed with cancer was described as traumatic. Weeping, fear, sadness, and extreme worry were the most common first reactions. Most women chose to disclose their condition only to their immediate family due to fears of being stigmatized. Ann, a hospital nurse, chose not to disclose her condition to her hospital colleagues and sought treatment at another hospital. She feared being only viewed as a cancer patient and not as a respected senior nurse.

Impact of Breast cancer

Breast cancer had differing impacts on the women's everyday activities, physical health, female identity, roles, and responsibilities. However, every woman reported having financial difficulties. Moreover, the women received misinformation and faced issues accessing medication. Breast cancer impacted the interviewees' physical health and everyday activities. Common effects of breast cancer and treatment included inability to eat, mood swings, lack of sexual desire, weight loss, pain, urinary incontinence, fatigue, and hair loss. Some women attributed their reduction in social interactions to physical issues and fear of stigmatization. Cancer impeded their ability to attend church and community events. Four women completely stopped working and felt having breast cancer hindered their ability to take care of their family.

> Without the silicone [breast] to support me, I feel like people will see it and say something.
>
> – Gifty

Conversely, several women affirmed that the disease did not, or only minimally, affect their daily living activities, family roles, and female identity. Most daily life disruptions were at the onset of the disease. With treatment, the condition of some women improved, and they resumed their usual activities. Jane and Talia tied their identity as women to fulfilling their wifely duties. Jane felt that her inability to work and provide financially put a strain on her marriage. Talia's husband provided financially but was emotionally withdrawn and did not support her having a mastectomy. Due to the intrinsic link to the female identity, physical change to the breast can have a profound psychological effect that may lead to treatment default. Lynn's fear of undergoing a mastectomy led her to seek alternative treatment, defaulting treatment for three months.

Accessing medicines elucidated geographical and financial barriers, engendering expenses and lost wages. Living outside Kumasi and Accra incurred additional transportation costs, with women traveling seven to ten hours by road every three weeks for treatment. In addition to the limited availability of cancer medicines throughout Ghana, interviewees highlighted that the National Health Insurance Scheme only covered breast cancer medications at accredited facilities. Hence, some women had to travel long distances to obtain medicines, as they were either not available or nearly ten times more expensive closer to home. All interviewees reported financial hardship. The inability to work, combined with the high treatment costs, occasionally resulted in treatment default. Mansah defaulted for five months. Financial hardship and misinformation played a role in Patricia delaying radiotherapy; she relied on an incorrect cost estimate provided by a friend as higher than what it was.

Regaining confidence

At this stage, women expressed confidence that they would beat the cancer. They regained confidence given the support they received and the success of their coping mechanisms.

> Initially, it wasn't easy but with the treatment and now I am feeling better, I can see some changes. I'm glad I am better as compared to when I started treatment.
>
> – Patricia

The women reported that their families provided some form of financial, physical, and/or emotional support. Talia's husband was financially supportive but did not provide any physical or emotional support. While it made her feel like a burden, Mary received physical and financial support from church members as they provided money and transportation for hospital visits.

Religion was the principal coping mechanism. Attending church activities, praying, and practicing gratitude made these women hopeful.

> I will always concentrate with God and the doctors. God will help them to cure my sickness.
>
> – Jane

Counseling by healthcare providers and breast cancer survivors also played an important role in decreasing fears and anxiety. Speaking with women who had breast surgery alleviated Gifty's anxiety. Overall, despite the barriers and complexities of the disease, the women reported being very hopeful and were determined to comply with medical recommendations.

Discussion

This study compared quality of life and psychological distress of women with breast cancer to controls, and explored the lived experiences of women with breast cancer. Cancer diagnosis, age, type of treatment, and education were significant predictors of quality of life and psychological distress. The lived experiences of women with cancer followed four major stages in the breast cancer journey: knowledge and suspicion, navigating the Ghanaian health system, impact of breast cancer, and regaining confidence.

When compared to the control group, breast cancer patients scored significantly lower on the physical health, psychological well-being, and environment domains. The women with cancer reported that pain and frequent hospital visits hindered their ability to conduct their daily living activities and to work. Regarding psychological well-being, women with breast cancer had more difficulty than the control group in accepting their bodily appearance, as weight loss and breast operations led to dramatic physical changes. Moreover, women with breast cancer perceived their lives as less meaningful than the control group. The women linked feeling unsafe to their uncertainty about the future. Additionally, accessing care took a heavy financial toll on the women living with breast cancer. Additionally, women with breast cancer also had fewer opportunities to take part in leisure activities when compared to the control group. The female breast is intrinsically linked to self-esteem, gender identity, femininity, and motherhood (Akin-Odanye et al., 2011). Hence, breast cancer can be a source of significant stress and may harm a woman's psychological well-being.

The women with breast cancer had significantly higher psychological distress and had more bouts of despair, anxiety, stress, and depression when compared to the control group. The higher psychological distress among the Ghanaian women living with breast cancer was due to experiencing intrinsic factors, such as intense physical health problems, as well as extrinsic factors, including difficulty navigating the healthcare system, lack of financial and social support, and other psychosocial stressors such as stigma, marriage difficulties, and loss of employment. Intrinsic and extrinsic factors can negatively affect psychological distress and predispose cancer patients to mental ill health (Berard et al., 1998).

Despite the statistically significant difference between women with breast cancer and women in the control, the actual score differences were small. On the quality-of-life domains, the difference ranged between 1.21 and 1.95 on a 20-point scale. While there was only a three-point difference on a 40-point scale for psychological distress, the average score was clinically significant as it placed women with breast cancer in the range of moderate mental disorder. Lastly, the results of the regressions within the breast cancer population must be interpreted cautiously as the sample is small and resulted in wide confidence intervals. Education, age, and type of treatment were significant correlates for quality of life and psychological distress within the breast cancer population. Those with higher levels of education scored higher for psychological well-being and self-reported quality of life. Educated individuals are more likely to have employment and higher levels of income, which are protective factors against the development of mental illness (Patel, 2001). Older women with breast cancer scored higher on the social relationship domain, and on self-reported quality of life. Type of treatment had an impact on the quality of life and the psychological distress of the women. Compared to cancer patients who did not receive the treatment, those who received radiotherapy scored higher on psychological well-being. However, these women scored higher for psychological distress. Moreover, patients who had radiotherapy and/or breast surgery were more satisfied with their quality of life and health than were those who did not receive these treatments. In contrast, women who received chemotherapy scored lower on self-reported quality of life than those who did not. An explanation for this result may be where they were in their breast cancer journey. Women receiving chemotherapy are usually in the early stages of the breast cancer journey and experiencing major physical impairments. In contrast, women who had breast surgery and/or radiotherapy are usually further along the journey and may be regaining their confidence through coping mechanisms, such as practicing religion and maintaining social interaction.

Our results are consistent with other African studies. Nigerian women with breast cancer had minimal to severe depressive symptoms, but women with breast cancer knowledge and higher education had lower levels of depression (Akin-Odanye et al., 2011). Adaptive coping mechanisms, such as practicing religion, has been shown to aid breast cancer patients to regain their confidence and improve quality of life and emotional and functional well-being (Kugbey et al., 2019; Gonzaga, 2013; Aziato & Clegg-Lamptey, 2015).

Our study points to potential interventions, including: reducing financial and geographical barriers to drug access, improving the hospital referral system, increasing the number of breast cancer awareness campaigns, increasing the availability of breast cancer screening, implementing a mandatory patient education by healthcare professionals, promoting adaptive coping mechanisms, and offering counseling services. National awareness campaigns can increase the population's knowledge of breast cancer and dispel disease-related myths while equipping women with the knowledge to navigate the Ghanaian health system and encourage hospital presentation (McKenzie et al., 2018). Through patient education and counseling, patients are provided with specific information on their condition, and can ask questions, which consequently ease feelings of uncertainty and anxiety.

Strengths and limitations

Recruiting 64 breast cancer patients and 64 controls was a major benefit, as it enabled the detection of a medium size effect between the two groups on the quantitative scales. Furthermore, utilizing a mixed-methods approach enabled us to achieve the objective of deepening our understanding of the lived experiences of breast cancer patients, and specifically the disease's impact on the lives of Ghanaian women. Bracketing and critical reflexivity enhanced the credibility of the findings. Another strength is the use of multiple forms of triangulation: source, method, and investigator. The qualitative interviews further confirmed the quantitative results. No contradictions were observed. Lastly, to address the language barrier, the scales and interview schedule were translated in Twi, and two translators assisted with data collection.

Despite using the back translation method, the meaning of certain question items may have been diminished, presenting a limitation. The fact that the controls were recruited from PLH and breast cancer screening events may have introduced bias into the study. Women who have high levels of psychological distress may choose to isolate themselves and not venture out to social activities, such as screening events. Conversely, women with high levels of psychological distress may feel compelled to go these events as they may be anxious about their health.

Conclusion

Women with breast cancer had lower scores than those in the control group for physical health, psychological well-being, environment, and psychological distress, and for overall satisfaction with health and quality of life. Cancer diagnosis did not have a significant effect on social relationships and self-reported quality of life.

Access to treatment was a major burden for breast cancer patients due to financial and geographical barriers. This was a major source of psychological distress. Increasing the availability and public insurance coverage of cancer medicines across Ghana's health institutions would enable equitable and universal access to affordable and efficacious breast cancer medicines (addressing SDG 3.b). Implementing universal health coverage (addressing SDG 3.8) that improves the availability and accessibility of breast screening and a drug delivery service could reduce geographical and financial barriers while increasing the likelihood that women with cancer stay in the workforce (Mensah & Mensah, 2020; Gbenonsi et al., 2021). Moreover, there is a need to implement more social protection programs (addressing SDG 1.3) that will allow patients to focus on healing from breast cancer without engendering the financial hardship that may drag their families into poverty (addressing SDG 1). There was a temporal element to psychological well-being among breast cancer patients, where those further down the breast cancer journey scored higher. Hence, there is a particular need to focus on supporting those early in the journey, especially to prevent treatment default. Despite improvement in psychological well-being, our results indicate that it is still important to provide women later in the journey with support, as they may still be experiencing psychological distress.

Patient education "requires that patients are well informed about their disorder and the resources available to them" (Distelhorst et al., 2015). Patients should be equipped with the ability to recognize and treat breast cancer-related physical and psychological side effects. While mindful of West African sociocultural beliefs, women should also be informed of their clinical stage and grade, and any information relevant to their condition (Distelhorst et al., 2015). Education of treatment benefits and financial support were also recommended to improve treatment access for SSA breast cancer patients by Foester and colleagues (2019). Mental health and counseling services should be readily available and accessible for these women.

Patient navigators, usually lay community members, "identify, anticipate, and help to alleviate any barrier a patient may encounter when seeking screening, diagnosis and treatment" (Freeman, 2006). Peer navigators were accepted among Tanzanian women (Koneru et al., 2017). Patient navigation has been effective in improving "adherence to breast cancer screening, follow-up of diagnostic abnormalities, the initiation of breast cancer treatment and stability or improvement in quality of life" (Robinson-White et al., 2010). The Helping Others through Personal Experience (H.O.P.E.) program at PLH involves training cancer survivors as peer navigators for patients. Such a program could be beneficial to women with breast cancer if expanded nationwide. Our study elucidated patient-centered interventions that could improve access to care, alleviating the psychological distress and improving the quality of life of breast cancer patients.

References

Adewuya, A. O., Ola, B. O., Aloba, O. O., Mapayi, B. M., & Okeniyi, J. A. (2008). Impact of postnatal depression on infants' growth in Nigeria. *Journal of Affective Disorders, 108*(1–2), 191–193.

Akin-Odanye, E. O., Asuzu, C. C., & Popoola, O. A. (2011). Measured effect of some socio-demographic factors on depression among breast cancer patients receiving chemotherapy in Lagos State University Teaching Hospital (LASUTH). *African Health Sciences, 11*(3).

Andrews, G., & Slade, T. (2001). Interpreting scores on the Kessler psychological distress scale (K10). *Australian and New Zealand Journal of Public Health, 25*(6), 494–497.

Aziato, L., & Clegg-Lamptey, J. N. A. (2015). Breast cancer diagnosis and factors influencing treatment decisions in Ghana. *Health care for women international, 36*(5), 543–557.

Berard, R. M. F., Boermeester, F., & Viljoen, G. (1998). Depressive disorders in an outpatient oncology setting: Prevalence, assessment, and management. *Psycho-Oncology, 7*(2), 112–120.

Bougie, E., Arim, R. G., Kohen, D. E., & Findlay, L. C. (2016). Validation of the 10-item Kessler psychological distress scale (K10) in the 2012 aboriginal peoples survey. *Health Reports, 27*(1), 3.

Bowden, A., & Fox-Rushby, J. A. (2003). A systematic and critical review of the process of translation and adaptation of generic health-related quality of life measures in Africa, Asia, Eastern Europe, the Middle East, South America. *Social Science & Medicine, 57*(7), 1289–1306.

Brakohiapa, E. K., Armah, G. E., Clegg-Lamptey, J. N. A., & Brakohiapa, W. O. (2013). Pattern of breast diseases in Accra: Review of mammography reports. *Ghana Medical Journal, 47*(3), 101–106.

Cohen, J. (1992). A power primer. *Psychological Bulletin, 112*(1), 155.

Clegg-Lamptey, J., Dakubo, J., & Attobra, Y. (2009). During treatment in Ghana? A pilot study. *Ghana Medical Journal, 43*(3).

Dedey, F., Wu, L., Ayettey, H., Sanuade, O. A., Akingbola, T. S., Hewlett, S. A., . . . & Adanu, R. (2016). Factors associated with waiting time for breast cancer treatment in a teaching hospital in Ghana. *Health Education & Behavior*. doi: 10.1177/1090198115620417

De Menil, V., Osei, A., Douptcheva, N., Hill, A. G., Yaro, P., & Aikins, A. D. G. (2012). Symptoms of common mental disorders and their correlates Among women in Accra, Ghana: A population based survey. *Ghana Medical Journal, 46*(2), 95–103.

Distelhorst, S. R., Cleary, J. F., Ganz, P. A., Bese, N., Camacho-Rodriguez, R., Cardoso, F., . . . & Anderson, B. O. (2015). Optimisation of the continuum of supportive and palliative care for patients with breast cancer in low-income and middle-income countries: Executive summary of the Breast Health Global Initiative, 2014. *The Lancet Oncology, 16*(3), e137–e147.

Fann, J. R., Thomas-Rich, A. M., Katon, W. J., Cowley, D., Pepping, M., McGregor, B. A., & Gralow, J. (2008). Major depression after breast cancer: A review of epidemiology and treatment. *General Hospital Psychiatry, 30*(2), 112–126.

Foester, N., Ng, C., & McAllister, N. (2019). Risk of breast cancer following detection of breast lesions of uncertain malignant potential: A 23-year retrospective review. *Clinical Radiology*, *74*, e1.

Freeman, H. P. (2006). Patient navigation: A community based strategy to reduce cancer disparities. *Journal of Urban Health*, *83*(2), 139–141.

Gbenonsi, G., Boucham, M., Belrhiti, Z., Nejjari, C., Huybrechts, I., & Khalis, M. (2021). Health system factors that influence treatment delay in women with breast cancer in sub-Saharan Africa: A systematic review. P 1–16. https://doi.org/10.21203/rs.3.rs-234691/v1

Giacomini, M. (2010). Theory matters in qualitative health research. In by Ivy Lynn Bourgeault, V. L., De Vries, R. & Dingwall, R. (Eds). *The SAGE handbook of qualitative methods in health research*, 125–156. Thousand Oaks, CA: Sage.

Gonzaga, M. A. (2013). Listening to the voices: An exploratory study of the experiences of women diagnosed and living with breast cancer in Uganda. *Pan African Medical Journal*, *16*(1).

Groenvold, M., Petersen, M. A., Idler, E., Bjorner, J. B., Fayers, P. M., & Mouridsen, H. T. (2007). Psychological distress and fatigue predicted recurrence and survival in primary breast cancer patients. *Breast Cancer Research and Treatment*, *105*(2), 209–219.

Harper, A., & WHOQOL Group. (1996). *WHOQOL-BREF: Introduction, administration, scoring and generic version of the assessment* (Rep.).

Hay, I. (2010), *Qualitative research methods in human geography*, 3rd edn. London, UK: Oxford University Press.

Igene, H. (2008). Global health inequalities and breast cancer: An impending public health problem for developing countries. *The Breast Journal*, *14*(5), 428–434.

Koneru, A., Jolly, P. E., Blakemore, S., McCree, R., Lisovicz, N. F., Aris, E. A., . . . & Mwaiselage, J. D. (2017). Acceptance of peer navigators to reduce barriers to cervical cancer screening and treatment among women with HIV infection in Tanzania. *International Journal of Gynecology & Obstetrics*, *138*(1), 53–61.

Kugbey, N., Asante, K. O., & Meyer-Weitz, A. (2019). Depression, anxiety and quality of life among women living with breast cancer in Ghana: Mediating roles of social support and religiosity. *Supportive Care in Cancer*, 1–8.

Kugbey, N., Meyer-Weitz, A., & Oppong Asante, K. (2019). Mental adjustment to cancer and quality of life among women living with breast cancer in Ghana. *The International Journal of Psychiatry in Medicine*, *54*(3), 217–230.

Kyei, K. A., Oswald, J. W., Njoku, A. U., Kyei, J. B., & Vanderpuye, V. (2020). Anxiety and depression among breast cancer patients undergoing treatment in Ghana. *African Journal of Biomedical Research*, *23*(2), 227–232.

McKenzie, F., Zietsman, A., Galukande, M., Anele, A., Adisa, C., Parham, G., . . . & McCormack, V. (2018). Drivers of advanced stage at breast cancer diagnosis in the multicountry African Breast Cancer-Disparities in Outcomes (ABC-DO) study. *International Journal of Cancer*, *142*(8), 1568–1579.

Mensah, K. B., & Mensah, A. B. B. (2020). Cancer control in Ghana: A narrative review in global context. *Heliyon*, *6*(8), e04564.

Miranda, J. J., & Patel, V. (2005). Achieving the millennium development goals: Does mental health play a role? *PLoS Med*, *2*(10), e291.

Morhason-Bello, I. O., Odedina, F., Rebbeck, T. R., Hardford, J., Dangou, J. M., Denny, L., & Adewole, I. F. (2013). Challenges and opportunities in cancer control in Africa: A perspective from the African Organisation for Research and Training in Cancer. *The Lancet Oncology*, *14*(4), 142–151.

Obrist, M., Osei-Bonsu, E., Awuah, B., Watanabe-Galloway, S., Merajver, S. D., Schmid, K., & Soliman, A. S. (2014). Factors related to incomplete treatment of breast cancer in Kumasi, Ghana. *The Breast, 23*(6), 821–828.

Ohene-Yeboah, M., & Adjei, E. (2012). Breast cancer in Kumasi, Ghana. *Ghana Medical Journal, 46*(1), 8–13.

Patel, V. (2001). Poverty, Inequality and mental health in developing countries. In: Leon, DA; Walt, G, (eds.) Poverty, Inequality and Health. London: Oxford University Press. pp. 247-262.

https://researchonline.lshtm.ac.uk/id/eprint/15373 Patel, V., & Kleinman, A. (2003). Poverty and common mental disorders in developing countries. *Bulletin of the World Health Organization, 81*, 609–615.

Prince, M., Patel, V., Saxena, S., Maj, M., Maselko, J., Phillips, M. R., & Rahman, A. (2007). No health without mental health. *The Lancet, 370*(9590), 859–877.

Richards, L., & Morse, J. M. (2007). *Readme first for a user's guide to qualitative methods* (Second edition). Thousand Oaks, CA: Sage Publications.

Robinson-White, S., Conroy, B., Slavish, K. H., & Rosenzweig, M. (2010). Patient navigation in breast cancer: A systematic review. *Cancer Mursing, 33*(2), 127–140.

Scherber, S., Soliman, A. S., Awuah, B., Osei-Bonsu, E., Adjei, E., Abantanga, F., & Merajver, S. D. (2014). Characterizing breast cancer treatment pathways in Kumasi, Ghana from onset of symptoms to final outcome: Outlook towards cancer control. *Breast Disease, 34*(4), 139–149.

Skeen, S., Lund, C., Kleintjes, S., Flisher, A., & MHaPP Research Programme Consortium. (2010). Meeting the millennium development goals in Sub-saharan Africa: What about mental health? *International Review of Psychiatry, 22*(6), 624–631.

Sung, H., Ferlay, J., Siegel, R. L., Laversanne, M., Soerjomataram, I., Jemal, A., & Bray, F. (2021). Global cancer statistics 2020: GLOBOCAN estimates of incidence and mortality worldwide for 36 cancers in 185 countries. *CA: A Cancer Journal for Clinicians, 71*(3), 209–249.

Sweetland, A. C., Oquendo, M. A., Sidat, M., Santos, P. F., Vermund, S. H., Duarte, C. S., . . . & Wainberg, M. L. (2014). Closing the mental health gap in low-income settings by building research capacity: Perspectives from Mozambique. *Annals of Global Health, 80*(2), 126–133.

Sylla, B. S., & Wild, C. P. (2012). A million Africans a year dying from cancer by 2030: What can research and control offer to the continent? *International Journal of Cancer, 130*(2), 245–250.

World Health Organization. (1996). *WHOQOL-BREF: Introduction, administration, scoring and generic version of the assessment: Field trial version, December 1996 (No. WHOQOL-BREF)*. Switzerland: World Health Organization.

Zainal, N., Hui, K., Hang, T., & Bustam, A. (2007). Prevalence of distress in cancer patients undergoing chemotherapy. *Asia-Pacific Journal of Clinical Oncology, 3*(4), 219–223.

Zelle, S. G., Nyarko, K. M., Bosu, W. K., Aikins, M., Niëns, L. M., Lauer, J. A., . . . & Baltussen, R. (2012). Costs, effects and cost-effectiveness of breast cancer control in Ghana. *Tropical Medicine & International Health, 17*(8), 1031–1043.

14 Event-history analysis of determinants of breastfeeding in Cambodia

Evidence from Demographic and Health Survey

Mengieng Ung and Sheila A. Boamah

Introduction

Breast milk is the universal standard for optimal infant nutrition and health and is one of the most important factors contributing to child survival, health, and growth (Scott *et al.*, 1999). Breast milk contains essential micronutrients and macronutrients (e.g., proteins, carbohydrates, and lipids) and energy required for infant growth and development during the first four to six months of life (Boix-Amorós *et al.*, 2019). It also provides various immunological and bioactive components that contribute to the child's immunologic defense system and increases their resistance to disease (Yang *et al.*, 2018). Optimal breastfeeding practices, according to the World Health Organization (WHO), consist of initiation of breastfeeding within the first hour of birth, exclusively for at least six months, and continuously for 24 months or beyond, to ensure long-term benefits for both the infant and the mother (World Health Organization, 2016). Studies have shown that improving breastfeeding practice could save over 820,000 lives of children under five years of age annually and about 20,000 annual deaths from breast cancer of the mothers (Victora *et al.*, 2016; Peñacoba and Catala, 2019). Exclusive breastfeeding has a significant effect on the incidence and mortality rate of infectious diseases in the short term, and in the long-term, reduces the risk of adult obesity, cardiovascular disease, type II diabetes, and other metabolic diseases (Abada, Trovato and Lalu, 2001; Singh and Singh, 2011; World Health Organization, 2017; Boix-Amorós *et al.*, 2019; Peñacoba and Catala, 2019).

In some developing countries, specifically in poor communities, breastfeeding not only is an appropriate recommendation for child health but it also ensures food security for infants and is considered a necessity for a child's survival (Rao and Kanade, 1992). Immensely in settings where water contamination is a significant issue and persistent threat to health, breastfeeding helps reduce children's exposure to harmful contaminants from drinking water and water-borne illnesses such as diarrhea or cholera (Arifeen *et al.*, 2001). In many developing countries, including Cambodia, the percentage of women who exclusively breastfeed during the first six months of life has risen; however, economic, socio-political, and cultural barriers affect a women's decision to breastfeed. For these reasons, exclusive

DOI: 10.4324/9780367743918-17

breastfeeding is at the forefront of the United Nations' Sustainable Development Goals (SDGs) and the Global Nutrition Targets to be achieved by 2030 to advance global child and adolescent health and achieve the under-five and neonatal mortality targets. The promotion of breastfeeding enables directly or indirectly the accomplishment of SDGs #2, 3, and 12. The proceeding sections will detail the historical accounts of breastfeeding practices in Cambodia, followed by the SDGs set to achieve a better and more sustainable future for new mothers and their children.

Historical accounts on breastfeeding initiatives

In the 1970s, the decline in breastfeeding became a global phenomenon due to the development of artificial baby milk and, by extension, an increase in supplementary feeding (Trussell *et al*., 1992; Koosha, Hashemifesharaki and Mousavinasab, 2008). To reverse the trend, the WHO and the United Nations Children's Fund (UNICEF) placed breastfeeding promotion as a high priority on their health policy agenda with a particular focus on developing countries (Trussell *et al*., 1992). In that light, the Baby-Friendly Hospital Initiative (BFHI) was launched in 1991 by the WHO and UNICEF following the Innocenti Declaration of 1990, aiming to protect, promote, and support breastfeeding (World Health Organization, 2017). The BFHI seeks to promote and support breastfeeding for the well-being of all mothers and their babies. Although BFHI was a progressive initiative, it could not address the need of some women aiming to improve infant feeding practice at the community level in developing country contexts, and therefore, the Baby-Friendly Community Initiative (BFCI), a new community initiative tailored to fit with developing country settings, was established (Health, 2009, 2016). Proven successful in other developing countries, the BFCI was first implemented in Cambodia in 2004. By the end of 2007, there were over 3,360 baby friendly communities/villages, expanding up to 20% of the village in 12 provinces in Cambodia. The number grew to 34% and 44% of baby-friendly villages in 22 provinces by 2011 and 2014, respectively (Chea, 2016). Owing to the BFCI, the rate of exclusive breastfeeding in Cambodia rose from 11% in 2000 to 74% in 2010 (Parks, 2013; Sanders, 2017). Although progress has been made, there have been reports of early termination of breastfeeding with increasing use of breastmilk substitutes in Cambodia (Prak *et al*., 2014).

Breastfeeding and the Sustainable Development Goals

In 2016, the SDGs by the 2030 agenda succeeded the MDGs by the 2015 agenda, and subsequently, nine more goals were added into the SDGs agenda to tackle issues including equity, climate change, and environment (Shingirai, Katsinde and Chandrasekhar, 2016). Of the 17 SDGs, three are directly linked to breastfeeding (Goals 2, 3, and 12):

A dedicated global goal, SDG2 is based on a comprehensive approach to tackling food insecurity and improving nutrition, especially among the vulnerable and the poor, and ending all forms of malnutrition leading to stunting and wasting of children under the age of five years by 2030, and promoting sustainable agriculture. At the global level, hunger and food insecurity are on the rise, and malnutrition still affects millions of children. However, exclusive breastfeeding is an effective strategy to prevent malnutrition as breastfeeding provides a vital source of nutrition for children, which can eliminate hunger for children especially, for those in poor communities (Rao and Kanade, 1992; Shingirai, Katsinde and Chandrasekhar, 2016).

Goal 3 is to ensure healthy lives and promote well-being with the targets of ending preventable deaths of newborns and children under five years old. In 2018 alone, about 5.3 million children died before reaching their fifth birthday; almost half of those deaths, or 2.5 million, occurred within the first 28 days of life (the neonatal period). By 2018, 121 countries met the SDG target on under-five mortality, and 21 countries are expected to do so by 2030, including Cambodia. Not only does breastfeeding help to achieve this goal, but it could also save babies' life from water-borne diseases and improve their overall health and well-being.

The third goal, SDG12, is about ensuring sustainable consumption and production and achieving economic growth. The focus is on decoupling economic growth from environmental degradation, increasing resource efficiency, and promoting sustainable lifestyles. Breastmilk production is a natural process, which does not require industry for production, packaging, transport, or storing. Unlike the high proportion of food that is often lost along the supply chain, breastfeeding and human milk is considered a sustainable consumption that produces a negligible ecological footprint.

The study

While a plethora of studies focuses on the benefits of breastfeeding, understanding of the determinants of the initiation, exclusive breastfeeding, and termination practices primarily in the context of Cambodia remains unclear (see Abada, Trovato and Lalu, 2001; Crookston *et al.*, 2007; Sasaki *et al.*, 2010; Wren and Chambers, 2010). The emphasis on factors that determine the duration beyond exclusive breastfeeding among mothers is crucial in advancing our understanding of breastfeeding pattern and termination practices, which are useful in informing policies that seek to promote breastfeeding in Cambodia and other developing countries, as a significant component of achieving SDGs 2, 3, and 12. The present study aims to fill this knowledge gap by examining timing (1 to 36 months) to terminate breastfeeding and associated factors in Cambodia.

Method

Participants

Participants in this study were drawn from the 2014 Cambodia Demographic and Health Survey (CDHS). The 2014 CDHS is a national-level population and health survey administered to people aged 15–49 years. It is the fourth comprehensive survey conducted in Cambodia as part of the worldwide MEASURE DHS project since 2000 (Cambodia Demomographic and Health Survey, 2014). A total of 5,453 analytical samples of women who delivered their last child in the past five years were included in the analysis. The focus on the very last child over the past five years is to minimize recalled bias.

Outcome variable

The primary outcome variable is the duration of breastfeeding of the last child. The variable was calculated as the total number of months the mother reported having breastfed their last child. The WHO and UNICEF (2003) recommended exclusively breastfeeding for the first six months of age and continue into the second year or longer. To capture the changes over time (from 1 to 36 months of breastfeeding), we censured those women who fell out of this time range (World Health Organization and UNICEF, 2003).

Explanatory variables

Explanatory variables were categorized into four different groups: socio-economic and demographics (SED), biological, psychosocial, and media. SED variables include age (in a 10-year grouping starting from 15–24 to 45+), respondent's education level (0 = no education, 1 = primary education, 2 = secondary education and 3 = higher), place of residence (0 = urban, 1 = rural), wealth index (from poorest = 1 to riches = 5), respondent's type of occupation (0 = not working, 1 = skilled job and 2 = labor job), wanted the last child born (0 = no, 1 = yes), and gender of last child (0 = male, 1 = female).

Breastfeeding is a complex human behavior. A growing body of research suggests that many factors influence breastfeeding practice across cultures. For instance, the literature suggests a strong and consistent association between mother's age and level of education with the duration of breastfeeding (Scott and Binns, 1999; Abada, Trovato and Lalu, 2001). Compared with younger women, older ones are more likely to initiate and adhere to breastfeeding for more extended periods (exclusive to continuous breastfeed) (Bulk-Bunschoten *et al.*, 2001; Gudnadottir *et al.*, 2006; Novotny et al., 2000). Therefore, the mother's age is chosen to be the core predictor of the breastfeeding period in the Kaplan Meier Survival Curves (KMSC) analysis (Figure 14.1). Gender bias in the duration of breastfeeding has also been reported, whereby in some communities in Northern Thailand, India, Bangladesh, and Egypt, baby boys are breastfed longer than baby girls (Brown *et al.*, 1982; Jackson *et al.*, 1992; Rao and Kanade, 1992; Nath and Goswami, 1997). On the other hand, in Kenya, Hungary, the UK, the USA, Australia, and Switzerland, baby girls are breastfed longer than boys (Margulis, Altmann and Ober, 1993; Bouvier and Rougemont, 1998; Scott *et al.*, 1999; Cronk, 2000; Dunbar, 2002).

Furthermore, biological factors such as parity, method of delivery, pre- and postnatal care have also been found to influence the duration of breastfeeding (Thulier

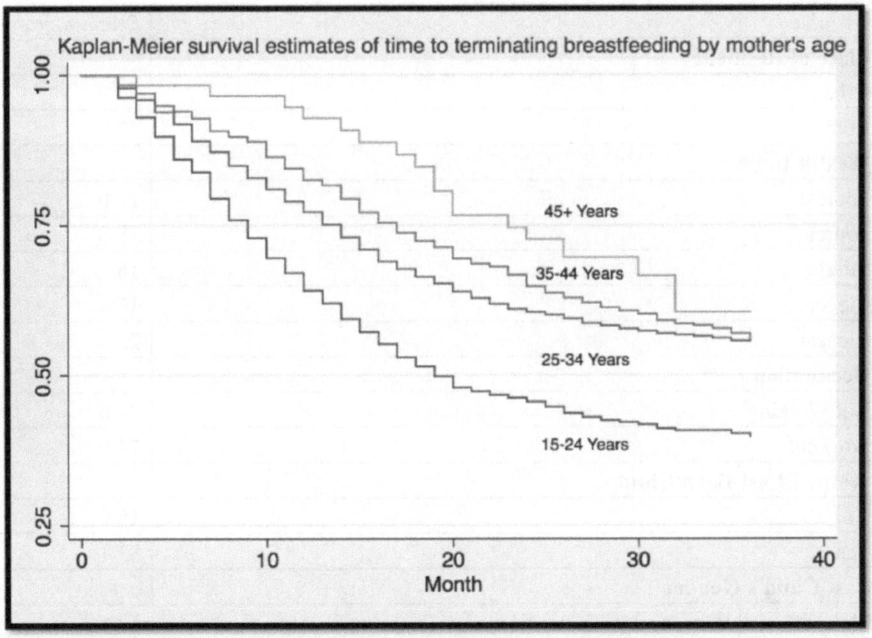

Figure 14.1 Kaplan-Meier estimates of time to terminating breastfeeding

Table 14.1 Descriptive analysis of outcome and predictor variables of time to terminate breastfeeding among married women in Cambodia (N = 5453)

Variable	*Percentage*
Outcome Variable	
Duration of Breastfeeding (Mean/Standard Deviation)	19.1 (11.67)
Predictor Variables	
Socio-economic and demographics	
Age	
15–24	30.7
25–34	55.2
35–44	12.9
45+	1.2
Education	
No Education	14.4
Primary	48.5
Secondary	33.2
Higher	3.9

(*Continued*)

Table 14.1 (Continued)

Variable	*Percentage*
Place of Residence	
Urban	26.7
Rural	73.3
Wealth Index	
Poorest	24.0
Poorer	19.9
Middle	16.2
Richer	17.4
Richest	23.5
Occupation	
Not Working	26.0
Working	74.0
Wanted Last Born Child	
No	16.9
Yes	83.1
Last Child's Gender	
Male	50.8
Female	40.2
Biological	
Parity	
1	32.7
2	33.4
3	16.8
4	8.7
5	4.1
6+	4.3
Method of Delivery	
Cesarean Section	6.8
Natural	93.2
Psychosocial	
Place of delivery	
Home	15.9
Public Settings	70.2
Private Settings	13.9
Prenatal Care with Doctor Ψ	–
Postnatal Care visit	
No	18.7
Yes	81.3
Media @	–

Source: The Cambodia Demographic and Health Surveys 2014

@: Variables were the outcome of Principle Component Analysis and Mercer, 2009). High parity is associated with shorter birth spacing and, therefore, a shorter duration of breastfeeding for each child (Abada, Trovato and Lalu, 2001). In this analysis, we included parity – the total number of ever born children (range from 1 to 6+) and method of delivery (0 = caesarean section, 1 = natural). Psychosocial factors included the place of delivery (0 = home, 1 = public settings including the health center, public hospital, and 2 = private settings, including private clinic, private hospital). We conceptualize prenatal care as a composite variable from prenatal care visits to a health care provider, including a doctor, nurse, midwife and traditional birth attendant, and village health volunteers. Postnatal care visit during the first six months of delivery was coded (0 = no, 1 = yes). Finally, exposure to information through media was also a composite variable created using the frequency of reading newspapers, listening to the radio, and watching television.

Analysis

We conducted descriptive and survival analyses in response to the study objectives. The descriptive analysis of the duration of breastfeeding and mother's age are presented using KMSC (Figure 14.1). Although KMSC estimates the hazard of the event and provides an estimated probability of surviving an event at any time point, it does not account for confounding or effect modification by other covariates (Vittinghoff *et al.*, 2011). Survival analysis was conducted to control for covariates to address these limitations.

The proportionality assumption did not hold after a log–log analysis comparing the baseline hazard of breastfeeding practice among mother's age. Therefore, parametric survival analysis was used with the assumption that the hazards across the different groups were non-proportional but rather exponential (Singer and Willett, 2003). The multivariate analyses were categorized into three models – Model I: socio-economic and demographics, Model II: biological, and Model II: psychosocial and media.

Results

In this section, we present univariate and multivariate relationships between the timing of termination of breastfeeding and sociodemographic factors, biological and psychosocial attributes of the participants. Table 14.1 illustrates the descriptive statistics of outcome and explanatory variables. The mean duration of breastfeeding for children was about 19 months. The majority of women were in 15–24 and 25–34 age groups. Forty-eight percent of women completed primary education, and 33% completed secondary education. About 73% of respondents resided in a rural setting. Twenty-four percent of women were in the poorest wealth category, 20% in a poorer category, 16% in the middle category, 17% in a richer category, and 23 % in the richest category (Table 14.1). Seventy-four percent of women were employed, and approximately 83% of women wanted their last-born child. About 51% of the last-born child was male. The majority of the women had at least two children. About 93% of women delivered their last child naturally. The place of delivery was 16%, 70%, and 14% for home, public health institutions, and private health institutions, respectively. About 81% had postnatal care visits.

In Figure 14.1, KMSC presents the respective survival curves for women in the 15–24, 25–34, 35–44, and 45+ age categories. The survival curves represent the probability that a woman will continue to breastfeed at any given time. During the first six months, age-specific differences in the cessation of breastfeeding by mothers are not discernible, especially for mothers in the 25–34 and 35–44 age groups. Noticeable trends were observed over time, with variable rates of decline in breastfeeding by age. For instance, at 20 months, the probability of women in the 45+ age group continuing breastfeeding is 0.80, whereas the probabilities of women in the 15–24, 25–34, and 35–44 age groups were 0.50, 0.67, and 0.72, respectively. The predictors of time to termination of breastfeeding are presented in Table 14.2.

Table 14.2 Hazard Ratios from multivariate hazard models of time to terminate breastfeeding among married women in Cambodia (N = 5,093)

Predictor Variables	*SED*		*Biological*		*Psychosocial and Media*	
	HR	*SE*	*HR*	*SE*	*HR*	*SE*
Socio-Economic and Demographics						
Age (Ref: 15–24)						
25–34	0.63***	0.03	0.64***	0.03	0.59***	0.03
35–44	0.58***	0.04	0.60***	0.06	0.45***	0.04
45+	0.42***	0.10	0.46***	0.11	0.28***	0.07
Education (No Education)						
Primary	1.03	0.07	1.02	0.07	0.99	0.07
Secondary	1.05	0.08	1.03	0.08	1.02	0.08
Higher	1.20	0.18	1.15	0.18	1.40*	0.22
Place of Residence (Ref: Urban)						
Rural	1.15**	0.07	1.15*	0.07	1.10	0.07
Wealth Index (Ref: Poorest)						
Poorer	0.93	0.06	0.93	0.06	0.90	0.06
Middle	0.89	0.06	0.88	0.06	0.87	0.06
Richer	0.97	0.07	0.96	0.07	0.89	0.07
Richest	0.80***	0.07	0.77***	0.07	0.72***	0.07
Occupation (Ref: Not Working)						
Working	0.55***	0.03	0.55***	0.02	0.50***	0.02
Wanted Last Born Child (Ref: No)						
Yes			1.10	0.06	0.95	0.06
Last Child's Gender (Ref: Male)						

Predictor Variables	*SED*		*Biological*		*Psychosocial and Media*	
	HR	*SE*	*HR*	*SE*	*HR*	*SE*
Female			0.97	0.04	0.99	0.04
Biological						
Parity (Ref:1)			0.98	0.02	1.08***	0.02
Method of Delivery (Ref: C-Section)						
Natural			0.80**	0.7	0.86	0.07
Psychosocial						
Place of delivery (Ref: Home)						
Public Settings					1.39***	0.09
Private Settings					1.22*	0.11
Prenatal Care with Doctor (Ref: No)						
Yes					1.06*	0.02
Postnatal Care visit (Ref: No)						
Yes					0.94	0.02
Media					0.97	0.02
Number of Failure	5088		5072		4276	
Log-Likelihood	−5086.76		−5064.93		−4537.09	

Source: The Cambodia Demographic and Health Surveys 2014

* P < 0.05, ** p < 0.01 and ***p < 0.001 HR = Hazard Ratios, SE = Standard Errors, SED (socio-economic and demographics)

In Model I, socio-economic and demographic variables were used. Mother's age was a significant determinant of time to terminate breastfeeding like in the KMSC survival curves (Table 14.2 and Figure 14.1). The older women had a lower probability of termination of breastfeeding than the youngest category of women. Women who reside in rural settings had a 15% higher probability of terminating breastfeeding. Women in the richest wealth quintile were less likely to terminate breastfeeding than individuals in the poorest category. Those employed maintained breastfeeding for more extended periods.

In Model II, we introduced biological covariates into the Model. The findings consistently showed that older women breastfed their children for a longer duration than younger women (Table 14.2), which is consistent with the KMSC result (Figure 14.1). Rural residents had a 15% higher probability of termination of breastfeeding compared to their urban counterparts. Compared to women in the poorest group, wealthier women had a lower probability (HR = 0.77, p < 0.001) to terminate breastfeeding. Women who were working had a lower probability of termination of breastfeeding, and those who delivered their last baby naturally breastfed their child longer when compared with women who delivered their child by the caesarean section (Figure 14.2).

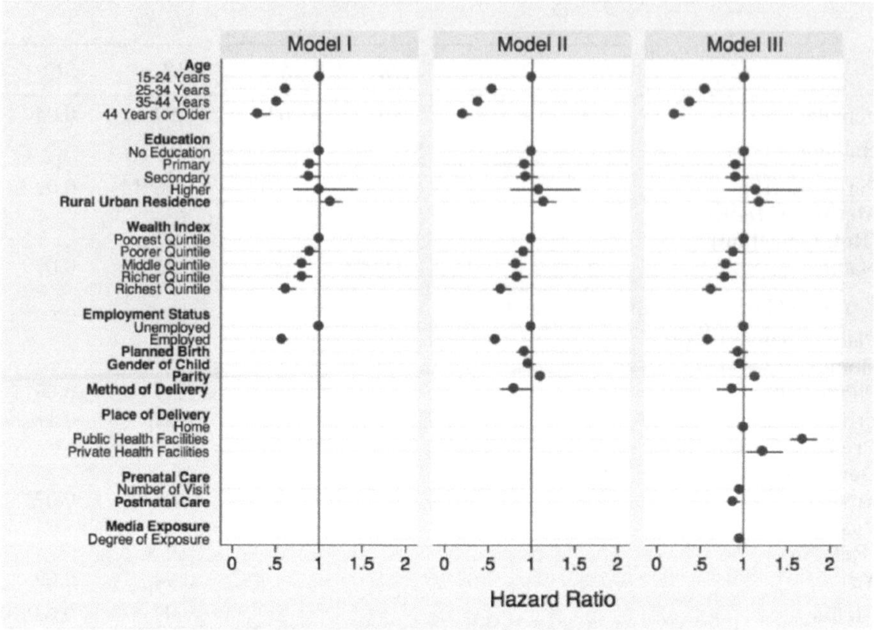

Figure 14.2 Graphical presentation of multivariate hazard models of time to terminate breastfeeding among married women in Cambodia (N = 5,093)

In Model III, we added Psychosocial and Media variables into the analysis. Once more, women's age, wealth, and employment maintained a significantly lower probability of termination of breastfeeding even after accounting for psychosocial and media covariates. Place of residence became insignificant after controlling for psychosocial related variables. The relationships between education and parity became significant after controlling for psychosocial attributes and access to media variables. Women who completed formal higher education were more likely to terminate breastfeeding early (HR = 1.40, $p < 0.05$), and so were those who had more than one child (HR = 1.08, $p < 0.001$) (Table 14.2). Compared to those who delivered their child at home, those delivered at public and private health institutions were more likely to cease breastfeeding early. Women who had prenatal care were more likely to terminate breastfeeding than their counterparts who did not receive prenatal counseling (Figure 14.2).

Discussion

Situated in the broader context of SDGs, this study examined time to terminate breastfeeding among married women in Cambodia using respondents' age as the main predictor while controlling for theoretically relevant covariates. Our findings indicate that older women breastfed their children longer compared to younger women. However, in the literature, the relationship between women's age and the duration of breastfeeding is inconclusive. Aryal (2007) found a negative relationship between age and the duration of breastfeeding, whereas Scott and colleagues found otherwise (Scott and Binns, 1999; Aryal, 2007). There are two possible explanations for these inconsistencies. On the one hand, it might be considered culturally inappropriate for older women to still breastfeed a child. On the other hand, due to fecundity, older women are less likely to become pregnant and, therefore, have time to breastfeed longer (Trussell *et al.*, 1992). Besides, older women might experience more breastfeeding role models (mother or grandmother) than younger women to emulate. They are more likely to reject modern breastmilk substitutes and rely on traditional forms of infant feeding, including prolonged breastfeeding.

Previous studies (Balogun *et al.*, 2017; Gao *et al.*, 2016; Senanayake *et al.*, 2019) indicate that breastfeeding prevalence systematically varies by urbanicity (rural-urban status). The literature suggests that typically urban women breastfeed for a shorter duration than their rural counterparts (Trussell *et al.*, 1992; Abada, Trovato and Lalu, 2001) Tang *et al.*, 2019; because, in urban areas, the adoption of bottle-feeding is considered to be more modern, sophisticated, and convenient primarily if the mother works outside the home. Our findings, on the other hand, are inconsistent with the foregoing studies. We found that women in an urban area instead breastfed longer than those are in a rural area under certain circumstances as it was not significant in the final Model. Although this finding may seem counterintuitive, it is still worth noting because it seems to suggest that there is an association between a mother's educational status and her likelihood of breastfeeding. The educational status of the mother is an important social determinant of health for children as it might impart an effect on health and infant feeding behavior (Acharya and Khanal, 2015). It is well-established that compared to their urban counterpart, rural women are less likely to be highly educated, have less wealth, and might not have adequate access to prenatal care services due to remoteness, which eventually influences the duration of breastfeeding. Our final model suggests that women with higher education were more likely to breastfeed longer period than women with lower educational attainment.

To end hunger (SDGs # 2) and to ensure healthy living (SDGs # 3), tackling rural–urban breastfeeding practice disparities is imperative. One of Goal 3 targets is to reduce neonatal and under-five mortality. Vanthy and others (2019) found that Cambodian children in rural areas have a 1.53 times greater risk of under-five mortality compared with their urban counterparts. Rural Cambodia is often a resource-poor setting, where access to clean water and food remains inadequate (Vanthy *et al.*, 2019). Breastfeeding is also important in achieving SDG 12, which is about ensuring sustainable consumption and production patterns. Breastmilk plays a critical role in preventing children's exposure to water-borne diseases, ensuring food security, and ultimately curbing under-five mortality. Breastmilk is naturally produced without packaging or processing needed, which creates a neglectable carbon footprint. Breastmilk is instantly ready whenever the babies need to feed; there is no storage or discard of leftover. Hence, breastfeeding reduces food and solid waste, as well as raise awareness of the importance of sustainable consumption.

We found that employed women breastfed their children longer, which is contrary to the literature (Lindberg, 1996; Ong *et al.*, 2005; Baker and Milligan, 2008; Cooklin, Donath and Amir, 2008). However, analyses of the distribution of the sample indicate that employed mothers were more clustered in the older group, were educated and in richer wealth quintile. For this reason, this category of women had adequate resources to enable them to adhere to breastfeeding. Given that the women in richer wealth quintiles were more likely to be nourished, it partly explains why they extend the duration for breastfeeding. On the contrary, mothers in the lowest wealth quintile ceased breastfeeding early; this might be due to poor nutritional status, which could potentially reduce the volume and quality of breast milk. Our findings further suggest that multiparous women (higher parity), who were predominantly rural dwellers, breastfed their last child shorter compared to primiparous women (lower parity). High parity is associated with a shorter birth interval and, therefore, shorter times available for breastfeeding.

Policy recommendation and conclusions

This study has demonstrated that time to cessation of breastfeeding is a complex and multifactorial phenomenon influenced by biological, psychosocial, sociodemographic, and other dynamics. This study provides a clearer picture of rural and urban disparities in time to termination of breastfeeding within Cambodia and has implications for developing countries with similar ethnic and urbanicity status. The study determined that women in rural areas are more likely to cease breastfeeding, warranting the need for developing breastfeeding interventions for rural communities, particularly targeting the poorest population. The use of age, rural/urban, and wealth categories provides a nuanced understanding of differences in breastfeeding behaviors among lactating mothers. These findings hold several policy implications. We suggest six areas to focus support in order to improve the duration of breastfeeding: mothers and their families, communities, health care, employment, research and surveillance, and public health infrastructure. In this context, information on appropriateness and benefits of breastfeeding should be emphasized through prenatal and postnatal care programs and through mass media. It is also pertinent to encourage the promotion of health care facility routines that will be conducive to breastfeeding, which includes rooming-in and demand feeding, immediate postnatal contact between mother and child. The benefit derived will be minimal if the duration of breastfeeding is short. Therefore, there is still an ongoing need for educational campaigns on proper breastfeeding methods and benefits of lactation, which should focus not solely on promoting initial and exclusive breastfeeding but the continuous breastfeeding. Improving breastfeeding practices not only provides a direct benefit to the children and mother but also contributes to achieving SDGs goals at large. Finally, policy on adopting breastfeeding-friendly in the workplace could address structural barriers and stigmatization toward new mothers, which could hinder women from achieving nonmaternal opportunities during the lactation period.

References

Abada, T. S. J., Trovato, F. and Lalu, N. (2001) 'Determinants of breastfeeding in the Philippines: A survival analysis', *Social Science & Medicine*, 52(1), pp. 71–81.

Acharya, P. and Khanal, V. (2015) 'The effect of mother's educational status on early initiation of breastfeeding: Further analysis of three consecutive Nepal Demographic and Health Surveys Global health', *BMC Public Health*, 15(1). doi: 10.1186/s12889-015-2405-y.

Arifeen, S. *et al.* (2001) 'Exclusive breastfeeding reduces acute respiratory infection and diarrhea deaths among infants in Dhaka slums', *Pediatrics*, 108(4), pp. e67–e67.

Aryal, T. R. (2007) 'Breastfeeding in Nepal: Patterns and determinants', *Journal of Nepal Medical Association*, 46(165).

Baker, M. and Milligan, K. (2008) 'Maternal employment, breastfeeding, and health: Evidence from maternity leave mandates', *Journal of Health Economics*, 27(4), pp. 871–887.

Balogun, O. O., Dagvadorj, A., Yourkavitch, J., da Silva Lopes, K., Suto, M., Takemoto, Y., . . . and Ota, E. (2017) 'Health facility staff training for improving breastfeeding outcome: A systematic review for step 2 of the Baby-Friendly Hospital initiative', *Breastfeeding Medicine*, 12(9), pp. 537–546.

Boix-Amorós, A. *et al.* (2019) 'Reviewing the evidence on breast milk composition and immunological outcomes', *Academic.Oup.Com*, 8(77), pp. 541–556. Available at: https://academic.oup.com/nutritionreviews/article-abstract/77/8/541/5492496 (Accessed: 14 March 2021).

Bouvier, P. and Rougemont, A. (1998) 'Breast-feeding in Geneva: Prevalence, duration and determinants', *Sozial-und Präventivmedizin*, 43(3), pp. 116–123.

Brown, K. H. *et al.* (1982) 'Consumption of foods and nutrients by weanlings in rural Bangladesh', *The American Journal of Clinical Nutrition*, 36(5), pp. 878–889.

Bulk-Bunschoten, A. M. W., Van Bodegom, S., Reerink, J. D., Jong, P. P. D. and De Groot, C. J. (2001) Reluctance to continue breastfeeding in The Netherlands. *Acta Paediatrica*, 90(9), pp. 1047–1053.

Cambodia Demomographic and Health Survey. (2014) *National Institute of Statistics and ICF Macro, 2014*. Available at: https://dhsprogram.com/pubs/pdf/fr312/fr312.pdf (Accessed: 14 March 2021)

Chea, M. (2016) *Complementary Feeding Intervention in Cambodia: BFCI. ILSI SEA Region Seminar on Maternal, Infant and Young Child Nutrition: Updates on Cambodia, Lao PDR and Myanmar*. Phnom Penh, Cambodia: ILSI SEA Region.

Cooklin, A. R., Donath, S. M. and Amir, L. H. (2008) 'Maternal employment and breastfeeding: Results from the longitudinal study of Australian children', *Acta Paediatrica*, 97(5), pp. 620–623.

Cronk, L. (2000) 'Female-biased parental investment and growth performance among the Mukogodo'. In L. Cronk, N. Chagnon, & W. Irons (Eds)., *Adaptation and Human Behavior: An Anthropological Perspective*. London: Routledgepp. 203–221.

Crookston, B. T. *et al.* (2007) 'Buddhist nuns on the move: An innovative approach to improving breastfeeding practices in Cambodia', *Maternal & Child Nutrition*, 3(1), pp. 10–24.

Dunbar, R. I. M. (2002) 'Helping-at-the-nest and sex-biased parental investment in a Hungarian gypsy population', *Current Anthropology*, 43(5), pp. 804–809.

Gao, H., Wang, Q., Hormann, E., Stuetz, W., Stiller, C., Biesalski, H. K. and Scherbaum, V. (2016) 'Breastfeeding practices on postnatal wards in urban and rural areas of the

Deyang region, Sichuan province of China', *International Breastfeeding Journal*, 11(1), pp. 1–11.

Gudnadottir, M., Sigurdur Gunnarsson, B. and Thorsdottir, I. (2006) 'Effects of sociodemographic factors on adherence to breastfeeding and other important infant dietary recommendations', *Acta paediatrica*, 95(4), pp. 419–424.

Health, C. M. of (2009) *Implementation Guidelines for Baby Friendly Community Initiative*. Phnom Penh, Cambodia: Ministry of Health Cambodia.

Health, K. M. of (2016) *Baby Friendly Community Initiative: Implementation Guidelines*. Nairobi, Kenya: Ministry of Health Kenya.

Jackson, D. A. *et al.* (1992) 'Weaning practices and breast-feeding duration in Northern Thailand', *British Journal of Nutrition*, 67(02), pp. 149–164.

Koosha, A., Hashemifesharaki, R. and Mousavinasab, N. (2008) 'Breast-feeding patterns and factors determining exclusive breast-feeding', *Singapore Medical Journal*, 49(12), p. 1002.

Lindberg, L. D. (1996) 'Women's decisions about breastfeeding and maternal employment', *Journal of Marriage and the Family*, pp. 239–251.

Margulis, S. W., Altmann, J. and Ober, C. (1993) 'Sex-biased lactational duration in a human population and its reproductive costs', *Behavioral Ecology and Sociobiology*, 32(1), pp. 41–45.

Nath, D. C. and Goswami, G. (1997) 'Determinants of breast-feeding patterns in an urban society of India', *Human Biology*, pp. 557–573.

Novotny, R., Hla, M. M., Kieffer, E. C., Park, C. B., Mor, J. and Thiele, M. (2000) 'Breastfeeding duration in a multiethnic population in Hawaii', *Birth*, 27(2), pp. 91–96.

Ong, G. *et al.* (2005) 'Impact of working status on breastfeeding in Singapore', *The European Journal of Public Health*, 15(4), pp. 424–430.

World Health Organization, W. H. (2016) *Infant and Young Child Feeding-Fact Sheet*. Available at: www.who.int.libaccess.lib.mcmaster.ca/mediacentre/factsheets/fs342/en/ (Accessed: 14 March 2021).

World Health Organization (2017) 'Guideline: Protecting, promoting and supporting breastfeeding in facilities providing maternity and newborn services'. Geneva: World Health Organization.

Parks, C. (2013) *Breastfeeding in Cambodia Is the New 'Normal'*. Available at: www.unicef.org/cambodia/12633_20202.html

Peñacoba, C. and Catala, P. (2019) 'Associations between breastfeeding and mother-infant relationships: A systematic review', *Breastfeeding Medicine*, Mary Ann Liebert Inc., pp. 616–629. doi: 10.1089/bfm.2019.0106

Prak, S. *et al.* (2014) 'Breastfeeding trends in Cambodia, and the increased use of breast-milk substitute: Why is it a danger?', *Nutrients*, 6(7), pp. 2920–2930.

Rao, S. and Kanade, A. N. (1992) 'Prolonged breast-feeding and malnutrition among rural Indian children below 3 years of age', *European Journal of Clinical Nutrition*, 46(3), pp. 187–195.

Sanders, J. G. (2017) 'Cambodia's breastfeeding rates high despite policy gaps', *Cambodia Daily*. Available at: www.cambodiadaily.com/news/cambodias-breastfeeding-rates-high-despite-policy-gaps-133342/

Sasaki, Y. *et al.* (2010) 'Predictors of exclusive breast-feeding in early infancy: A survey report from Phnom Penh, Cambodia', *Journal of Pediatric Nursing*, 25(6), pp. 463–469.

Scott, D. A. *et al.* (1999) 'Factors associated with the duration of breastfeeding amongst women in Perth: Australia', *Acta Paediatrica*, 88(4), pp. 416–421.

Scott, J. A. and Binns, C. W. (1999) 'Factors associated with the initiation and duration of breastfeeding: A review of the literature', *Breastfeeding Review: Professional Publication of the Nursing Mothers' Association of Australia*, 7(1), pp. 5–16.

Senanayake, P., O'Connor, E. and Ogbo, F. A. (2019) 'National and rural-urban prevalence and determinants of early initiation of breastfeeding in India', *BMC Public Health*, 19(1), pp. 1–13.

Shingirai, M., Katsinde, S. and Chandrasekhar, S. (2016) 'Breast feeding and the sustainable development agenda', *Indian Journal of Pharmacy Practice*, 9. doi: 10.5530/ijopp.9.3.2

Singer, J. D. and Willett, J. B. (2003) *Applied Longitudinal Data Analysis: Modeling Change and Event Occurrence*. Oxford: Oxford University Press.

Singh, N. S. and Singh, N. S. (2011) 'Determinants of duration of breastfeeding amongst women in Manipur', *Bangladesh Journal of Medical Science*, 10(4), pp. 235–239.

Tang, K., Gerling, K., Chen, W. and Geurts, L. (2019) 'Information and communication systems to tackle barriers to breastfeeding: Systematic search and review', *Journal of Medical Internet Research*, 21(9), e13947.

Thulier, D. and Mercer, J. (2009) 'Variables associated with breastfeeding duration', *Journal of Obstetric, Gynecologic, & Neonatal Nursing*, 38(3), pp. 259–268.

Trussell, J. *et al.* (1992) 'Trends and differentials in breastfeeding behaviour: Evidence from the WFS and DHS', *Population Studies*, 46(2), pp. 285–307.

Vanthy, L. *et al.* (2019) 'Determinants of children under-five mortality in Cambodia: Analysis of the 2010 and 2014 demographic and health survey', *Int Arch Public Health Community Med*, 3, p. 29. doi: 10.23937/2643-4512/1710029

Victora, C. G. *et al.* (2016) 'Breastfeeding in the 21st century: Epidemiology, mechanisms, and lifelong effect', *The Lancet*, 387(10017), pp. 475–490.

Vittinghoff, E. *et al.* (2011) *Regression Methods in Biostatistics: Linear, Logistic, Survival, and Repeated Measures Models*. New York, NY: Springer Science & Business Media.

World Health Organization and UNICEF (2003) *Global Strategy for Infant and Young Child Feeding*. Geneva: World Health Organization.

Wren, H. and Chambers, L. (2010) 'Breastfeeding in Cambodia: Mother knowledge, attitudes and practices', *World Health & Population*, 13(1), pp. 17–29.

Yang, T. *et al.* (2018) 'Nutritional composition of breast milk in Chinese women: A systematic review', *Asia Pac J Clin Nutr*, 27(3), pp. 491–502. doi: 10.6133/apjcn.042017.13

15 The world we want

The development we want

Andrea Rishworth and Cristina D'Alessandro

Gender, health, and sustainable development are inextricably linked. As Langer et al. (2015, 1168) remarked, "when women are valued, enabled and empowered, gender equality and health can be achieved; and when women are healthy and have equity in all aspects of life, sustainable development will be possible." The preceding chapters have not only gone a considerable way toward extending these connections, but have also demonstrated how gender, health, and sustainability are inherently geographical. As many contributors are geographers, space, scale, and time are all implicit themes. This is evident in underlying contextual and compositional factors that shape one's access to and use of resources that vary over time, and converge when examining the gender-health-development nexus at scales from personal decisions and behaviors, to global organizations and their political arrangements. The dynamic and layered landscapes of gender permeate our entire world.

These complex influences on gender realities reveal that permeate gender realities reveal that both "gender" and "development" can too easily be reduced to simplistic definitions. While development and feminism share philosophies of transformation, competing political objectives lead to struggles for interpretive power that commonly essentialize particular notions of women and serve to legitimize problematic interventions. We need to embrace the complexity inherent in gender and development, for as Cornwall et al. (2007) remind us, it is precisely when the context-specific nature of gender relations is forgotten, that gender myths are created, internalized, and upheld. Gender is more than a synonym for women; rather, it is a multidimensional, dynamic, and relational concept comprising the norms, behaviors, roles, and relations of humans that vary across space and change over time. Development too must be understood as more than Gross National Product, and assumptions of linear progression generated through productive power; rather, it is the process of expanding and enhancing the freedoms of life (Sen, 1999). Recognizing how and why gender and development are inherently intertwined is thus a critical challenge, because power and representation have profound implications on people's lives worldwide.

While there are many encouraging examples of development initiatives promoting gender equality and sustainable development, we cannot presume these interconnections are always self-evident. Few examples reflect this potential

DOI: 10.4324/9780367743918-18

disconnection between development attempts for gender equality than the gender mainstreaming approaches that followed the 1995 Beijing Conference on Women. These gender mainstreaming ideologies in the late 1990s were underpinned by the notion that policies and programs would put women's and men's concerns as central to the achievement of gender equality, yet in so doing, simply "added women and stirred" into existing development plans. As a result, gender mainstreaming created simplified concepts of gender inequality, where women are seen as passive recipients of development assistance (Kumari, 2013). This is all too clear in many African countries where gender mainstreaming initiatives undertaken by development organizations have created a host of unintended consequences – rape, violence, suffering – due to culturally insensitive, blanket agendas that mask the geographical nuances of people and place (Standing, 2007; Wendoh & Wallace, 2005). As Lawson, Anfaara, and Osman (Chapter 3) contend, competing ontologies of development contradict notions of gender equality due to global-local fissures and, in so doing, remind scholars that multidimensional forms of gender (in)equality must be understood, interpreted, and enacted within the contextual landscape in which they are found.

The complexities and uncertainties of attempting to promote gender equality and development are manifold, and according to Connell (2012; 1680), are not dependent on individualized gender realities, but rather "are always dynamic, and always interwoven with the dynamics of the world." Gender equality and development are shaped and embedded within multidimensional economic, affective, and symbolic power relations interacting at the interpersonal, intrapersonal, institutional, and societal levels. A half a century after Beijing Conference, Derry, Anchor, and Bisung's (Chapter 9) analysis of WASH services in Ghana reveals an example of the continuing effects of external development agendas on the oversimplification of gender dynamics and disparities in community settings. In the case of WASH, moving beyond a homogenous understanding of gender realities might involve asking how potentially different gender and development outcomes might have been with household and community engagement. Could a conversation around gender relations and responsibilities surrounding WASH produce synergistic effects for women and girls to attain gender equality, health and well-being? Paradoxically, perhaps, interventions led by the various communities that are disenfranchised have the best chances of success as "hopeful geographies" (Lawson, 2005). As Alhadeff and Mosely revealed in Chapter 10, not only is food (in)security deeply interconnected with gender, poverty, and politics but, so too, community engagement and innovations are inherently connected to the likely success of food and gender equality. The promotion of gender and health supporting policies for sustainable development thus constitutes a prudent, though challenging way forward.

Nevertheless, the ways to meet gender equality in sustainable development are inevitably taxing due to inherent issues of complexity – a theme consistently borne out in this volume. Baada and Baruh (Chapter 11) contend that adverse situations, such as underutilized antenatal care, are simply not a matter of health systems, but vary according to social and economic capital from place to place.

These dynamics are complicated further by intersectional identities, understood as the interconnected nature of social categorizations that create overlapping and interdependent systems of disadvantage. When these intersectional identities are left unexamined, women are often treated as a homogeneous group with shared experiences and, as a result, perpetuate inequality and the othering of groups into marginalized and disenfranchised positions and statuses (Crenshaw, 1990). As Ung and Boamah (Chapter 15) discussed, opportunities for health, through breast feeding for example, are differentially experienced among various segments of the female population; this enables some to experience good health, while others to suffer. The effects, however, are not solely individual but rather multidimensional, impacting families and communities to societies across generations.

A particular challenge to unravel gender-development complexity is the synergistic and co-acting effects gender and health inequality create for sustainable development. Boateng pointed out these intricacies in Chapter 8 revealing how the co-occurring inequalities in food, water, and energy tend to worsen health outcomes and opportunities for development at both the individual and societal scales. Syndemic interactions perpetuate mutually enhancing cycles of gender inequality and structural difficulties that co-occur in particular temporal and geographical contexts while interacting at the individual and societal levels. These interactions therefore make finding the roots of injustice vague and difficult to ameliorate. As Fernández, de la Torre, and Vallarta highlight in Chapter 6, the multidimensional nature of inequalities, together with their interactions with a host of social, ecological, and political processes, requires systems thinking approaches to yield solutions relevant at the local, regional, national and international scale. It is only when policies and practices reflect the multilayered contextual landscapes of people and place, that gender equality will be realized and acted upon.

The multidimensional causes, consequences, and implications of unequal gender and development opportunities are a consistent theme throughout the chapters, that despite laying bare stark inequalities, continue to persist across time and over space. Kangmennaang, Achore, Kalsi, Atiim, and Bisung remind readers in Chapter 12 that although women's access to and control over resources promote the social, cultural, political, and individual determinants that support women's empowerment and reduce of maternal mortality, the inability of governments and institutions to address these interconnections perpetuates gender, health, and development disparities. That is, while health is an intrinsic common good, a view echoed by the WHO's Commission on the Social Determinants of Health (2008), the intersectional interactions between various determinants are often neglected in favor of magic bullet solutions to gender and development. As a result, these magic bullet solutions fail to appreciate and examine how gender matters globally in the context of existing gender inequities. While knowledge and action on the roots and realities of gender, health, and development inequalities have come a long way since the Beijing Platform, renewed global attention is required if society is to reach the interrelated targets set for in the Sustainable Development Goals. Looking to the future, uncertain sustainable development realities pervade. The preceding chapters, however, provide hopeful strategies to overcome systemic barriers by tackling interrelated inequalities and generating innovative solutions. Yet, if success is to be met, we might consider the value of three additional interconnected themes.

First, attention to the role of leadership and commitment is urgently required in order to address *gender equality and the empowerment of women and girls* and recognize its indivisibility with all other sustainable development goals. Leadership from all levels of government and society is critically needed not just to address the unfinished gender agenda set forth in the MDGs, but to equally recognize the interconnected SDG goals and targets necessary to counter the persistent inequalities that global populations endure. Readers see this clearly in Chapter 2 where Owusu highlights the continued educational inequalities girls and women suffer, despite the pertinence of education in the cycle of empowerment and freedoms. As the United Nations explains, without coordinated responses and overlapping effects, universal access to education for girls and women, as articulated in both the MDGs and SDGs, will not be realized. A geographic lens is critically important because each level of government requires different forms of leadership, commitment, and responses. As Boateng and Williams in Chapter 13 remind us, policy interventions for health issues, such as breast cancer, not only require national purchase, but must also be tailored to the daily social, physical, and psychological challenges associated with health. This is supported by Nussbaum (2020), who states that while policy provision is necessary for the recognition of rights, policy provision alone is insufficient to ensure that individuals are able to convert policies into basic human functioning.

Second, appropriately addressing the sustainable development goals and targets requires a data revolution and novel statistical collection. This means tackling the digital divide within and between nations, and in societies across all scales through innovative methods such as real-time data collection, machine learning, and deep learning. This will facilitate synergies and decision-making power, and thus allow citizens to be active participants in their quest for a better life. As Osaki-Tomita, Gotoh, Ishitsuka, and Hojyo (Chapter 5) contend, without capturing gender statistics at both the micro and macro scales, systemic gender inequalities will persist and accentuate societal divides. Rudimentary, novel, and underexamined data sets are urgently needed to address persistent, growing, and new gender divisions, as well as expose often veiled gender inequities related to care and domestic responsibilities. As Akbari and Williams (Chapter 7) so clearly indicate, when a researcher is able to capture data relevant to the seemingly mundane aspects of people's everyday lives, such as care, researchers are capable of illuminating underexamined and often hidden gender inequalities and thus take requisite steps to address gender concerns. Without untangling the "web of causation" and moving beyond our reliance on particular data sources tied to particular problems, we will fail to integrate our knowledge, acquire new information, and reveal changing populations patterns and experiences of inequality.

Third, the complexity and interrelated nature of the SDGs requires multilayered policy implementation and consistent evaluation in order reach the 2030 goals. Policies must be crafted within, between, and across societal sectors, ministries, and institutions to provide guidance, consistency, and accountability that transform society. We see this need particularly in Chapter 4, where Ferroukhi, López, and Baruah reveal that without contextually relevant and applicable policies that meet the needs of women and their expert capabilities, strategies to alleviate gender inequalities remain obsolete. Policy, while imperative for designating rules of compliance, will fall flat without consistent evaluation across all spaces and scales of society. Strategies for policy evaluation will not only provide opportunities for learning, accountability, and procedures to operationalize the SDGs, but also generate new evidence that works with, and assesses the implementation of the SDGs in strategic and innovative ways. No more clearly can we see the need for tailored policy responses and consistent evaluation than in the case of the unfinished universal access to reproductive health and human rights MDG agenda that began over two decades ago.

Obviously, much work remains, not only to reach the few targets set forth in the MDGs, but also to attain, strive, and thrive under the SDGs. Yet, given the countless forms of gender inequality revealed in the preceding chapters, it seems unlikely that progress will be met under new policy prescriptions and government agendas. Social, economic, political, cultural, and environmental processes continue to entrench inequalities, inflict disparities, and re-shape forms of suffering in new and complicated ways. We see this plainly during the global COVID-19 pandemic, where 500,000 more girls are at risk of being forced into child marriage during the first year of the pandemic alone (i.e., Cousins, 2020), and one million more expected to become pregnant, fuelled by school closures and economic destitution (Cousins, 2020). In India, for instance, long and hard lockdowns left millions of daily laborers and migrant workers without any work and, as a result, pushed millions more people into poverty. In tandem, India's economy contracted almost a quarter, as schools remained closed and COVID-19 cases continued to skyrocket. This environment is now pushing many families to consider child marriage in order to alleviate their poverty and suffering. While the true extent of these blatant and changing gender inequalities remain uncertain due to imprecise data, this example does show how various forms of gender inequality are interconnected with dynamics embedded in the household, community, national and international scale.

In places of North America, such as the United States for example, the COVID-19 pandemic has also created new forms of gender suffering. While stay-at home orders enacted in many states were intended to protect the public and prevent widespread COVID-19 infection, new stay at home orders left victims of intimate partner violence trapped with their abusers. Although domestic violence hotlines prepared for an increase in demand for services, many places across the country experienced their call rate drop by over 50% (Fielding, 2020; Evans et al., 2020). While on face value these figures may suggest improvement in rates of intimate partner violence, there is a darker side to the story. Emerging studies suggest that new policy implemenations are creating many unintended consequences whereby victims are not only unable to safely connect with desperately needed support services but are further forced to experience retrenched inequalities in the social determinants of health, magnifying the point that when crises strike, hardships are differentially experienced and oftentimes exacerbate pre-existing inequalities (Kofman & Garfin, 2020; Agüero, 2021; Nix & Richards, 2021). These gender calamities and misfortunes, while persistent worldwide, do present opportunities for transformation and change. As contributors have meticulously illustrated, gender (in)equality and sustainable development are intrinsically interconnected. (Un)sustainable development processes often intensify gender (in)equalities through social, economic, political, and ecological pathways that unfold in particular places; so too gender (in)equalities can intensify (un)sustainable development through natural resources, land use, and water management practices. It is only by recognizing these interlocking dynamics across different spaces, in varied places and across scales, will gender inequality in all its guises be identified, addressed, and omitted. Arguably, the emergence of sustainable gender geographies presents a treacherous terrain to examine. Yet, for us to fully appreciate why gender matters globally requires that we start our investigations, for if we deny this challenge, matters of fundamental human rights and inequities in the prerequisites for meeting sustainable development goals around the world will continue to endure. When reflecting on this book's themes, we are left to ask what do we address first? Does gender equality spur sustainable development or does sustainable development lead to gender equality? Perhaps eliminating this dichotomous thinking and considering sustainable gender equality or gender sustainability, offer more productive alternatives for a healthy future. Rethinking these relationships globally thus offers an opening to address all sustainable development goals in a robust and comprehensive manner.

References

Agüero, J. M. (2021). COVID-19 and the rise of intimate partner violence. *World Development, 137*, p. 105217.

Connell, R. (2012). Gender, health and theory: Conceptualizing the issue, in local and world perspective. *Social Science & Medicine, 74*(11), pp. 1675–1683.

Cornwall, A., Harrison, E. and Whitehead, A. (2007). Gender myths and feminist fables: The struggle for interpretive power in gender and development. *Development and Change, 38*(1), pp. 1–20.

Cousins, S. (2020). 2·5 million more child marriages due to COVID-19 pandemic. *The Lancet, 396*(10257), p. 1059.

Crenshaw, K. (1990). Mapping the margins: Intersectionality, identity politics, and violence against women of color. *Stanford Law Review, 43*, p. 1241.

Evans, M. L., Lindauer, M. and Farrell, M. E. (2020). A pandemic within a pandemic: Intimate partner violence during Covid-19. *New England Journal of Medicine, 383*(24), pp. 2302–2304.

Fielding, S. (2020, April 3). In quarantine with an abuser: Surge in domestic violence reports linked to coronavirus. *The Guardian*. www.theguardian.com/us-news/2020/apr/03/coronavirus-quarantine-abuse-domestic-violence. opens in new tab.

Kofman, Y. B. and Garfin, D. R. (2020). Home is not always a haven: The domestic violence crisis amid the COVID-19 pandemic. *Psychological Trauma: Theory, Research, Practice, and Policy*, *12*(S1), p. S199.

Kumari, S. (2013). Gender and development: Exploring the successes and limitations of gender mainstreaming as a strategy. In *XXVII IUSSP International Population Conference to Be held in Busan, Republic of Korea, from August* (Vol. 26).

Langer, A., Meleis, A., Knaul, F. M., Atun, R., Aran, M., Arreola-Ornelas, H., Bhutta, Z. A., Binagwaho, A., Bonita, R., Caglia, J. M. and Claeson, M. (2015). Women and health: The key for sustainable development. *The Lancet*, *386*(9999), pp. 1165–1210.

Lawson, V. (2005). Hopeful geographies: Imagining ethical alternatives: A commentary on JK Gibson-Graham's "surplus possibilities: Postdevelopment and community economies". *Singapore Journal of Tropical Geography*, *26*(1), pp. 36–38.

Nix, J. and Richards, T. N. (2021). The immediate and long-term effects of COVID-19 stay-at-home orders on domestic violence calls for service across six US jurisdictions. *Police Practice and Research*, pp. 1–9.

Nussbaum, M. (2020). The capabilities approach and the history of philosophy. In Chiappero-Martinetti, Osmani and Qizilbash (eds.), *The Cambridge Handbook of the Capability Approach* (pp. 13–39). Cambridge, UK: Cambridge University Press.

Sen, A. (1999). *Commodities and Capabilities*. OUP Catalogue. Oxford: Oxford University Press, number 9780195650389.

Standing, H. (2007). Gender, myth and fable: The perils of mainstreaming in sector bureaucracies. In A. Cornwall, E. Harrison and A. Whitehead (eds.) *Feminisms in Development: Contradictions, Contestations and Challenges* (pp. 101–111). London, UK: ZED Books.

Wendoh, S. and Wallace, T. (2005). Re-thinking gender mainstreaming in African NGOs and communities. In F. Porter and C. Sweetman (eds.) *Mainstreaming Gender in Development: A Critical Review* (pp. 70–79). Phillidephia, PA: Oxfam GB.

World Health Organization and Commission on Social Determinants of Health (2008). *Closing the Gap in a Generation: Health Equity through Action on the Social Determinants of Health: Final Report of the Commission on Social Determinants of Health*. Geneva: World Health Organization.

Index

Note: Page numbers in *italics* indicate a figure and page numbers in **bold** indicate a table on the corresponding page. Page numbers followed by "n" indicate a note.